THE IMPORTANCE OF HAVING A BRAIN

BRAIN

TALES FROM THE HISTORY OF MEDICINE

THE IMPORTANCE OF HAVING A BRAIN

TALES FROM THE HISTORY OF MEDICINE

EXPANDED SECOND EDITION

TONSE N. K. RAJU

AUCTOREM
HOUSE

Auctorem House
276 5th Ave, Ste 704-2591
New York, NY 10001
www.auctoremhouse.com
Phone: 1 888-332-7718

Published by Auctorem House: 09/16/2025

ISBN: 978-1-965687-86-4(sc)
ISBN: 978-1-965687-87-1(e)

Library of Congress Control Number: 2025905431

Contents

Relief and Remedies

For Babies and Their Moms

Incredible Experiments

Mortals and Martyrs

Miscellaneous

Bibliography

To
VikramKamdar, MD
You continue to show us, by example,
how to be good doctors.

Preface

The medical profession is arguably the oldest of all professions.

Take midwifery and childbirth: Lions and elephants in labor don't summon friends for help or call their obstetricians. They move away from their herd to secluded locations to avoid predators at an extremely vulnerable period in their lives.

But humans almost always need assistance from at least one other person for safe childbirth. The evolution of upright walking (bipedalism) has led to a very narrow pelvic outlet, such that without assistance, childbirth would be very risky to both the mother and her baby. So, midwifery is an evolutionary necessity. Without it, our species probably would have long perished. The medical profession is therefore as old as human society.

Because of its extraordinary longevity, the history of medicine, too, is long and complex. It is a mixture of tales of amazing discoveries and inventions, as well as startling misadventures and utter stupidity. So, it is our story.

Like all histories, the growth of modern medicine has been a tortuous, if fascinating, path. Yet, such a wonderful story is seldom taught with any depth in medical schools. The medical, nursing, and dental students would be lucky to receive even a superficial and cursory introduction to medical history. When taught, it is often treated as borderline curiosity. Today's doctors are more likely to be adept at billing for an antibiotic treatment than knowing how the same antibiotic was discovered. They may know about DNA methylation but not about how the DNA molecular structure was deciphered.

My own introduction to the history of medicine was accidental. While a medical student, I edited the school magazine, *The Pulse*. For the magazine's special graduation edition, our surgery professor submitted an essay titled, *A Brief History of Doctors' Fee*. He had narrated the story of the *Code of Hammurabi*, the famous stone edict that detailed, among other things, a fee structure for surgical services based on patients' social status and the ability to pay. It also included punishments for poor results based on the patient's status and the extent of damage to the body. So, here was the first royally sanctioned medical fee structure with a built-in malpractice code issued several thousand years ago. My professor's insightful essay was funny, and thrilling to read.

The same professor then took me to a dusty corner of our library and showed me a small stack of medical history books that no one, except for him, had ever checked out.

"From today, *both* of us will be checking these out," he said. I took his advice and was hooked.

This book, started those many years ago in that medical school library's dusty corner, is my humble attempt to pay back my professor's gift to me. Over the past 40 years of my professional career, I have been writing these and similar stories in slightly varied forms in different venues. Printed in a variety of outlets—from free medical college newspapers, hospital cafeteria flyers, newsletters to professional societies, medical journals, textbook chapters, and the medical monthly *Hippocrates*—the stories have entertained and informed a variety of readers throughout the years.

The events, stories, and anecdotes retold here vary in length and depth, with no chronological sequence. The format is not accidental. I hope you find them fun and informative, rather than in-depth analyses. These are medical history stories dispensed in small doses for the novice.

Though the book is small, I owe a great deal of debt in bringing it out. I owe my gratitude to:

- A large number of primary and secondary source materials, of which I have cited only a few. The sources are acknowledged, but all errors in my text are mine. If readers point them out, I will correct them in future editions.
- My colleague and friend Caroline Signore, MD, who edited the text with a golden brush and immense kindness, and to Sharat Raju, who further cleaned up the manuscript.
- Arthur Balthazar, a Chicago-based cartoonist for his lively, if light-hearted, illustrations.
- Hundreds of my current and former students—my greatest teachers.
- Outskirts Press for production support.
- Vidya, Sharat, Manu, Archana and Valarie—for listening to these and other tales from me for so many years, most often willingly.

Tonse N. K. Raju, MD
February 2012
Gaithersburg, MD

Preface to the second expanded edition

The First Books Library published the first edition of this volume in 2002, and the Outskirts Press reprinted it in 2012. In this second, expanded edition, there are following changes.

- I have added new pictures and some extra text to the chapter "The Bellybutton Mystery" (pages 37—44).
- I have added 3 additional chapters in this edition: they are
  - o **Reconstructing memories and history in *One***

Hundred Years of Solitude **by Gabriel García Márquez.**

- o **Dr. Charles Alexander Eastman**
- o **Battling poverty, injustice, ignorance and fear, and despair**

- A slightly different version of the chapter "The Bellybutton Mystery", and the above three articles appeared in *Hektoen International* (volumes and numbers are acknowledged after the chapters). My thanks to the editors for permitting the use of the above text and figures from their periodical in my book.
- I thank Ms. Cait C. Ryan for the author photograph in the back cover of this volume.
- My appreciations are also due to Auctorem House for their publication efforts.

Tonse N.K. Raju, MD
May, 2025
Gaithersburg, Maryland, USA.

Smallpox Died in Somalia

On October 26, 1977, smallpox died in Somalia.

For all its might and fury, the smallpox virus that had terrorized mankind since antiquity dried out on the crusting skin lesions of a Somali resident, Mr. Ali Maow Maalin—the last human to contract natural smallpox.

Smallpox was born in Africa, only to die there some three thousand years later. The virus evolved over those three millennia and relentlessly spread across the globe, annihilating kingdoms and destroying civilizations. By the mid 18th century, no country had been spared from its wrath.

In the 20th century, more than 300 million died from smallpox— three times the number killed in World War I, World War II, and all wars and battles of the 20th century combined. There has never been a cure for smallpox, nor an effective method to stop its spread. Smallpox was feared and worshiped in many cultures.

Ancient Indians, Chinese, and Egyptians practiced inducing immunity from this disease by injecting smallpox pus obtained from dying victims into healthy subjects. The procedure, now called "variolation," made its way to the West through intriguing routes.

Lady Mary Wortley Montague, the wife of the English ambassador to Turkey, learned variolation while in Turkey, and secretly variolated her six-year-old son. She brought the practice to England. Before getting royal approval for its implementation, Princess Caroline wanted it to be tested for safety.

The princess commanded six condemned prisoners to undergo variolation with a promise of clemency for compliance, should they survive the procedure. All six agreed: they had little choice.

After variolation all six survived. The technique was declared safe and the prisoners were released. This classic case of poor ethics is also a case of poor science. All six volunteers in the royal experiment probably had smallpox in their childhood, and thus were immune from developing smallpox from variolation. The concept of lifelong immunity from natural smallpox infection had been known vaguely at the time, but not with any certainty.

Variolation gradually spread in the Old World, but remained dangerous, since the procedure often produced lethal smallpox. Edward Jenner (1749-1823), the 18th century English doctor, changed it all.

A smallpox epidemic in 1778 triggered Jenner's memory of a folk tale of cowboys and dairymaids who were previously infected with cowpox and were immune to smallpox. He thought about it for years and then in 1793, he concluded that for cowpox to protect against smallpox, a period of time must elapse between these infections. But he had no clue about the duration of that interval.

Sarah Nelmes, a local milkmaid, developed cowpox sores on her wrist in May 1796, and sought treatment from Jenner. On the 14th of that month, Jenner obtained consent for an ingenious experiment from the parents of James Phipps, a boy with no known history of smallpox or cowpox.

Jenner extracted a small amount of pus from Sarah's cowpox lesions and deposited it over scratches he had made earlier on the boy's arms. One week later the boy developed the expected cowpox symptoms and fully recovered.

Jenner then conducted one of the most daring experiments in medical history. On July 1, 1796, nearly two months after James's cowpox infection, Jenner inoculated the boy again—but this time using the more dangerous smallpox pus. Jenner waited to see what would happen. As Jenner predicted, James did not develop smallpox.

Edward Jenner's Experiment

The English doctor repeated the experiment on James a few more times and proved that James was now immune from smallpox—"Vaccination" was born. (Though, the term "vaccination" wasn't coined until nearly one hundred years later by Louis Pasteur. In 1881 he christened the procedure "vaccination," from *vacca* meaning cow in Latin, honoring Jenner in recognition of the cowpox, or "vaccinia virus" he used to prevent smallpox.)

News of Jenner's inoculation was looked upon with suspicion

by doctors, and the public. London newspapers lampooned Jenner, drawing caricatures of the doctor's face with a cow's body.

But, it did not take long for everyone to see that immunity from using Jenner's method of vaccination was real. Jenner's fame spread beyond English borders—even enemies of England honored him. In 1804, a beautiful Napoleon Medal was struck in his honor in France. The emperor's esteem of Jenner was so high that upon Jenner's requests, on many occasions he released English prisoners held in France. On one such occasion when Napoleon was about to reject a petition for clemency for two prisoners, Josephine reminded him that the clemency request came from Jenner. The emperor paused for an instant and reportedly exclaimed, "Jenner! Ah, we can refuse nothing to that man."

In a letter dated May 14th 1806, on the tenth anniversary of Jenner's inoculation of James, President Thomas Jefferson sent a letter to Dr. Jenner stating that he, the president, had been "among the early converts," and had recommended the procedure to his countrymen. He complimented Jenner for having "erased from the calendar of human afflictions one of its greatest."

After centuries of destruction, the closing chapters on the human battle against smallpox were written in the past century. In 1968, after much bickering and wrangling, the World Health Organization (WHO) invested a modest $2.4 million for the eradication of smallpox. The strategy was to achieve 80% global vaccination, timely case finding, and an effective surveillance system tagged onto bench research.

Ingenious methods were used to combat this disease. In India, for instance, 100 rupees were awarded for the informant who identified a new patient with smallpox. Once confirmed, hundreds of workers would turn up at that location and vaccinate everyone in all the villages nearby.

Over 200,000 workers were recruited to replicate this approach in countries where smallpox was endemic. This strategy worked. In ten years the hunt for smallpox was complete and so was its eradication. By the end of the 20th century, the WHO recommended discontinuing routine smallpox vaccinations.

Today, smallpox virus colonies exist only in the freezers at the Centers for Disease Control and Prevention in the United States and the State Research Center of Virology and Biotechnology in Russia. Scientists at the World Health Assembly of the WHO continue to debate about the definitive date for the complete destruction of these colonies. Some argue that colonies of live smallpox viruses should be maintained for research to develop safer, new-generation smallpox vaccines, manufacture new antiviral medications, and develop animal models for smallpox research.

These issues also stem from the concerns that accidental dissemination of smallpox virus cannot be ruled out, and worse still, there is always a risk of terrorists intentionally releasing the virus and causing a worldwide smallpox epidemic.

Now there is no need to fear the smallpox virus—we need to fear only one species of animals called the *Homo sapiens sapiens*—the wisest of all god's creatures.

While the story of the last remaining colonies of smallpox virus lives on, in 1980, four years after it had infected Ali Maow Maalin in Somalia, the WHO pronounced that the naturally occurring smallpox that had been born in Africa, in fact, had died in Africa.

Boston's Boylston Street

Boston has influenced the course of U.S. history like few other cities in this country. Birthplace of the nation, Boston showcases its past glory through scores of monuments and memorials. One of Boston's lesser known medical historical memorials, however, is Boylston Street.

Why is it named Boylston Street and who was Boylston?

A smallpox epidemic in 1721 afflicted half of Boston's 12,000 inhabitants and killed more than 900. The city's famous theologian Reverend Cotton Mather (1663-1728) had learned from one of his slaves, Onesimus, that in Africa "variolation" was practiced to prevent smallpox.

Mather also had read about the procedure in a science periodical, *Transactions,* which he received as a member of the London Royal Society. Variolation was *not* done using cowpox pus—that was to come 75 years later through Edward Jenner's monumental discovery in England. Variolation involved making small wounds on the skin of healthy persons and infecting them with a small amount of pus from other smallpox victims.

The Chinese, Indians, and Middle Easterners had been trying "inoculation" against smallpox since antiquity, by intentionally inducing milder forms to prevent severe smallpox. The technique was haphazard at best and the results were largely unpredictable.

The practice involved obtaining pus or powdered crusts from healed smallpox lesions and inserting the resulting substance into the skin with pins or other poking devices, or by inhaling it. This is the exact technique Lady Mary Wortley Montague, the wife of the British Ambassador to Turkey, used to have her son inoculated in Istanbul in 1717.

Back in Boston and desperate to do something to halt the epidemic, Mather tried to persuade Boston doctors to use variolation. No doctor agreed, except for Dr. Zabdiel Boylston (1679-1766).

On Monday, June 26, 1721, Boylston conducted variolation on his only son, Thomas, and on two slaves. All three developed a mild case of smallpox, but all recovered within a week.

The first smallpox inoculation in the American colonies was a success.

As the news of Boylston's experiment spread, a bitter controversy arose. Bostonians feared that inoculation would lead to the spread of smallpox instead of preventing it. Many members of the medical profession and the clergy strenuously opposed the inoculation practice and forbade Boylston from continuing the experiment.

A letter in a Boston newspaper was typical of the general reaction:

"...for a man to infect a family in the morning with smallpox and to pray to God in the evening against the disease is blasphemy;"

"...[that smallpox is] a judgment of God on the sins of the people... [and to] avert it is but to provoke Him more;"

"...[that inoculation is] an encroachment on the prerogatives of Jehovah, whose right it is to wound and smite."

The opposition was so severe that there were calls for trying Boylston for murder. Facing death threats and encountering attacks on the streets, the doctor stayed indoors after sunset. Grenades were thrown at the houses of Boylston and Mather. Mobs with nooses hunted them, intending to hang both men.

In spite of such fierce opposition, however, hundreds of Bostonians sought out Boylston for inoculation. Although intimidated by death threats, Boylston continued to perform the procedure. In a dramatic instance, he inoculated five members of the family of Mr. Edward Dorr. Four additional Dorr family members became ill, all at the same time. But all nine recovered, proving Boylston's critics wrong.

Boylston's cumulative experience was impressive: only 6 of the 244 (2.5%) inoculated inhabitants died, compared to 844 of the 5980 (14%) who had contracted the disease naturally.

It took one year for the epidemic to abate, and several more decades for Bostonians to appreciate the contributions of Cotton Mather and Zabdiel Boylston.

Mather and Boylston were far ahead of their times: Boylston carried out variolation 75 years before Jenner introduced an improved method of vaccination using cowpox; about 150 years before Louis Pasteur introduced the Germ Theory, proving that tiny microbes were capable of causing human illnesses; about 160 years before the first disease-causing viruses (in tobacco plants and in cattle) were discovered; and nearly 180 years before the basics of modern immunology were understood.

Eventually Bostonians recognized Zabdiel Boylston's contributions and honored him. In 1821, on the first centenary of the introduction of smallpox inoculation in the United States, they renamed Boston's Frogg Lane "Boylston Street," a major thoroughfare that runs more than 30 miles.

Boston's Boylston Street

The Large Pox Menace

Many infections cause skin lesions, or "boils." Swellings containing pus-like liquid, these lesions can vary in size, shape, and distribution. These features, along with the history of their onset and progression, give clues to doctors for diagnosing the infection. Finding a specific microbe responsible for the disease in the lesion or in patient's blood clinches the diagnosis and allows the doctor to set a course of treatment.

However, before scientists knew that microbes could cause disease, skin conditions were being described solely based on how they appeared, with a suffix *pox* attached to them.

Of all the "poxes," the syphilitic boils were the most dreadful. They were large, red, and painful, spreading rapidly over the body. Thus, syphilitic boils were called *large* or *great pox*. And so, smallpox, another dreaded disease, got the name "small" because the skin boils of smallpox were smaller in comparison to those of syphilis.

A few years after Columbus had returned from the New World, a major epidemic of syphilis broke out in Naples and in the rest of Italy from 1495-96. The epidemic spread quickly to France, England, Spain, Poland, Russia, and other European countries. Based on its presumed origin from the Spaniards, the Flemish and the Dutch called syphilis the *Spanish Sickness,* and the Portuguese, to be more specific, called it *Catalan Sickness.*

Soon there were other names, depending upon which country one disliked most. The French called it the *Naples (or Italian) sickness*, the Muscovites, *Polish sickness*, and the Poles, *German sickness*. The Germans, never having been fond of the French, called it the *French sickness*, with which the English and the Italians wholeheartedly agreed. Oddly, the Spaniards did not attribute the disease to any European nation; they assumed that it was an American malaise.

As syphilis spread to Asia, the Japanese and the East Indians called it the *Portuguese sickness* since the Portuguese were the first modern Europeans to colonize the Far East. In India, even today the reference to the "foreign" origin of syphilis persists: All venereal diseases are collectively named as *firangi (foreign)* afflictions.

What about the name "syphilis"? A Veronese physician, poet, and astronomer, Girolamo Fracastoro (1478-1553) wrote an elegant poem, *Syphilis: or a Poetical History of the French Disease*, in which the hero, a young shepherd by the name of Syphilis, was punished by Apollo for indiscretions. Given the venereal nature of its spreading, the name stuck.

To establish the origin of syphilis, researchers studied the skeletons from ancient populations using modern DNA technology. Bone lesions were found similar to those caused by syphilitic parasites from the remains of Americans dating back many centuries before Columbus's arrival. But no such evidence of syphilitic bone lesions have been found in the pre-Columbian European bone remains.

So, now a partial, if complex version has been proposed to explain the early origin and spread of syphilis: The parasite causing syphilis, *TrepenomaPallidum,* probably evolved over several millenniums, from a closely related but different trepenoma through mutation. Based on archeological evidence from bone pathology in a *Homo Erectus* in Kenya, scientists speculate that such a common ancestor might have evolved in East Africa.

But, how exactly it got out of Africa is a mystery yet to be solved.

Seeing with New Eyes

On January 29, 1680, the Royal Society of London elected a relatively under-trained amateur, a non-British gentleman to boot, as the "Fellow of the Royal Society" (FRS), the highest honor bestowed upon a scientist.

The Society's unprecedented action was well reasoned, for Antoni van Leeuwenhoek (1632-1723) was no ordinary scientist. He had discovered the universe of bacteria, protozoa and similar creatures and objects, never seen by any other pair of human eyes.

Leeuwenhoek was the son of a basket maker in Delft, Holland. Having served as an apprentice under a linen maker in Amsterdam, he began a successful drapery business. He also held many of the city's civic duties. He was a licensed surveyor, Delft city's official wine taster, and a chamberlain to the sheriff.

We do not know where, why, or how Leeuwenhoek learned the craft of lens grinding, nor did he write about it. Huddled in a closet of his house, he made hundreds of lenses and mounted them on 2- to 3-inch silver or brass tubing, creating high-precision microscopes with short focal lengths. He used innovative contraptions for illumination and managed to see and count small objects through his microscope, which surpassed all similar pieces of the time. At his death there were 247 microscopes and 172 lenses—some with magnifications of up to 300 times. He probably crafted several hundred more that have been lost.

Although Leeuwenhoek did not learn another language or attend a university, he was far from uneducated. As a bookkeeper he probably taught himself the skills needed to construct microscopes and delicate instruments.

Leeuwenhoek probably began assembling microscopes and using them early in his life. Using his instruments with skill and logical deductions, he made eye-opening discoveries. He studied water from rain, ponds, and wells, and examined human saliva and excreta in his microscopes. He described them in brief letters and papers.

Leeuwenhoek's first paper in 1673 was titled *A Specimen of Some Observations made by a Microscope Contrived by Mr. Leeuwenhoek, concerning Mould upon the Skin; Flesh etc; the Sting of a Bee etc.* He sent it to the Royal Society with a cover letter from his friend Dr. Reinier de Graaf. Despite its long and windy title and crude style, the scientists at the Society published the paper the same year in *Philosophical Transactions.* They ultimately published 374 more of Leeuwenhoek's papers.

Within three years, he had discovered the spermatozoa from insects, dogs, and man, the anatomy of the plant stem, algae, protozoa, and microscopic nematodes, the structure of woods and crystals, and human red blood cells.

He described what he saw in pithy, unsophisticated Dutch. For instance, he said in his September 17, 1683 letter to the Royal Society:

"...There are more animals living in the scum on the teeth in a man's mouth than there are men in a whole kingdom."

In 1674, Leeuwenhoek likely became the first human to see a microbial organism. Fascinated by their shape, their slender features, and their dancing undulations seen through the microscope, he called them "little animalcules" or small creatures.

Leeuwenhoek's discoveries became sensational during his lifetime. Royal visitors, such as Peter I the Great of Russia, James II of England, and Frederick II the Great of Prussia, came to see him and learn how he had developed microscopes. But their efforts were in vain; Leeuwenhoek did not show anyone the secret of making lenses or microscopes. His trade secret remained with him until his death on August 26, 1723 at the age of 90.

Leeuwenhoek was not the inventor of the microscope or the discoverer of cells. It was another Dutchman, Zacharias Jansen, a spectacle maker, who invented the microscope in 1590, which Galileo improved in 1610. And Robert Hooke (1635-1703) of the Royal Society of London published *Micrographia* in 1665, in which he described "cells" as the basic structures and subunits in the fragments of a wooden cork he had seen under a microscope.

Yet, Leeuwenhoek is credited for improving the microscope and for showing how to use it to study the universe of microbes. For the next two hundred years, scientists learned to appreciate the impact of his discoveries. Today, Leeuwenhoek is looked upon as the founder of modern bacteriology and microbiology, which evolved into cellular and molecular biology in the 20th and 21st centuries.

Leeuwenhoek bequeathed a huge collection of his microscopes to the Royal Society, most of which mysteriously disappeared upon his death in 1723. His daughter auctioned off the remaining items. Some of these are in display in the Amsterdam museum.

From a Country Doctor's Kitchen

On July 6, 1660, London was the site of a spectacular royal ceremony.

One month after his return from 17 years of exile, King Charles II restored a 700-year-old custom. His Majesty stroked and caressed the faces and cheeks of hundreds of his subjects ailing from *scrofula*, lymph node tuberculosis, and gave them a piece of royal gold to mark the occasion. With that, the royal chaplains proclaimed the *King's Evil* had been cured, and the grateful patients departed with hopes in their hearts and gold pieces in their pockets.

Tuberculosis had been one of the most dreaded diseases man had ever known. Its symptoms and signs had been described, but its cause remained elusive, and its treatment nonexistent. Known as the "Captain of Men of all Death," TB ravaged empires and humbled monarchs.

Unlike bubonic plague, which killed thousands at a time, tuberculosis lingered on, destroying one by one, gnawing away at the victim's vitality. While the plague was the *Black Death*, tuberculosis was the *White Plague*. Robert Louis Stevenson, Franz Kafka, Thomas Mann, and George Orwell were among its famous victims.

On March 24, 1882 in Berlin, Robert Koch (1843-1910), a German doctor, made a startling announcement. He had discovered the microbe that caused tuberculosis. It was a tiny, rod-shaped bacterium present in the lungs and sputum of TB patients. He could transmit the disease to animals by injecting the bacteria, and could recover the same bacteria from the infected animals. Koch became an instant celebrity.

The son of a mining officer, Koch was born in Klausthal, Germany. As a child, he wished to become an explorer and dreamt of winning the Iron Cross by working as a military surgeon. He loved the outdoors and studied nature by himself.

After a medical degree from the University of Göttingen he served briefly in the military during the Franco-Prussian War and later worked as a district medical officer in Wollstein (now in Poland). He had a busy practice and a lovely wife, Emma Fraatz. But with an interest in all things scientific, Koch pursued research in archaeology, anthropology, photography, and in the new field of bacteriology studying algae, fungi, and parasites.

On his 28th birthday Emma gave Koch a microscope as a gift so that he could play with it. She knew her husband's curiosity for scientific exploration. The gift changed Koch's life.

He mastered the fine art of adjusting light and focusing lenses. He experimented with staining methods. Reminiscent of Leeuwenhoek, Koch tried to see just about "everything under the sun" using his microscope. Part of the kitchen became his home laboratory.

One day, when he examined a drop of blood from a cow dying from anthrax, he saw among the blood cells numerous objects that looked like silk threads. To test if these "threads" were real, Koch studied tissues from healthy and sick animals, and from human victims of anthrax. He soon found the same silky creatures in humans, and deduced that this bacterium caused anthrax—the first microbe to be identified for causing a specific infection. By 1876, he had elucidated the complete life cycle of the anthrax bacillus.

In 1880, Koch took a research job at the Imperial Board of Health in Berlin. With two assistants, Koch searched for bacterial causes of other diseases and managed to transmit tuberculosis to laboratory animals using lung tissues from patients dying from rapidly progressive tuberculosis.

Initially, he could not see the tuberculosis bacterium on his glass slides. But, when he improved his staining technique by heating the glass slide with a smear of the sputum containing TB bacterium, and then stained the slide with a blue dye, he saw clumps of curved blue microbes. Thus, he was the first human to "see" the bacteria that caused tuberculosis.

Koch was not done: working in India and Africa he discovered microbes causing cholera, and did basic research on *trypanosomiasis*, malaria, and plague. He proposed the basic rules for proving an association between the causative microbe and specific infection, later called the *Koch's Postulates* in vogue even today. Its tenets are: The microbe should be found in patients suffering from the disease and not in those without the disease; when injected into laboratory animals, the disease should be transmitted; and the same microbes should be recovered from the infected animals.

Many of Koch's discoveries would have been worthy of Nobel Prizes. He was given the 1905 prize for the discovery of tuberculosis bacterium.

To commemorate Koch's astounding announcement of the discovery of the TB bacterium on March 24, 1882, the World Health Organization proclaimed March 24 of each year *World Tuberculosis Day*.

Although *Royal Touch* has disappeared, *King's Evil* still exists. Tuberculosis remains a global endemic, particularly in HIV-positive populations.

Doctor's Pub

A short walk from London's Piccadilly Circus subway station to the intersection of Broadwick and Lexington leads to a pub. In spite of the constant roar of motor vehicles, trams, and the buzz of tourists and office workers, it is a clean and nice place today.

This area, the Soho District in London, was not always so clean and nice.

In the mid 1800s, it teamed with poor people seeking employment. The place was filthy, smelly, with no sanitary services: Many home basements had cesspools of human waste underneath their floorboards. Since the cesspools were overrunning, the London city government decided to dump the waste into the River Thames. The contaminated water led to a major outbreak of cholera in August 1854.

Within three days, 127 people living on or near Broad Street (today's Broadwick Street) had died, and nearly three-quarters of inhabitants left their homes. By six months into the epidemic, 616 people had perished. Although not the first time cholera broke out in London, this was the most vicious epidemic to hit England.

John Snow (1813-58), the London doctor of obstetric anesthesia fame, was also an expert in epidemiology with an interest in the study of cholera.

During the 1854 London epidemic, Snow analyzed the geography of the water supply and mortality patterns in Soho. A disproportionate number of cholera cases occurred in regions supplied by one water company, with nearly five hundred cases within a few blocks of a single water pump on Broad Street.

This pump drew water from the Thames and a nearby well. A few feet away, Snow found a sewer pipe and reasoned that the pipe was contaminating the well and the pump water. He urged the authorities to "remove the handle from the pump." After some protest his advice was followed and the epidemic promptly abated.

Over the years since John Snow's death, London had grown, and with urbanization new buildings and restaurants came to Soho, including a pub on Broad Street.

In 1954, a group of doctors and medical historians requested the pub owners to display a plaque honoring the centenary of John Snow's research into the causes of the cholera epidemic of 1854. The owners obliged and even changed the pub's name to *John Snow*.

The pub's new sign was unveiled by Sir Austin Bradford Hill (1897-1991), then president of the Section of Epidemiology and Preventive Medicine of the Royal Society of Medicine.

The pub and the surroundings have been renovated, for it is one of London's historic sites. Inside *John Snow*, there are memorabilia commemorating Snow's discoveries.

While toasting Dr. Snow one should reflect that the shy and quiet doctor—the man responsible for curbing the great cholera epidemic of 1854—was a strict teetotaler.

Doctor's Pub: John Snow

All in God's Name

Stephen Hales (1677-1761), an English country pastor, measured blood pressure in living animals for the first time—but no one knew the importance of blood pressure at that time.

Hales was born in Bekesbourne in Kent and studied divinity at Cambridge. An ordained priest (1703) and a curate (1709) at the Teddington church, Hales was not trained in science. But guided by a sharp mind and unlimited curiosity, he performed scientific investigations with utmost concentration and devotion, as if he were in pursuit of God. Also, Hales might have been influenced by his friend and fellow Cambridge alumnus Isaac Newton, not to mention a uniquely creative, scientific atmosphere of the age.

To answer seemingly simple questions, Hales invented elegant, equally simple instruments, and designed experiments not unlike those today's physiologists might implement. With brilliant common sense and impeccable logic, he interpreted the results of his studies and came to rational conclusions. For things that he did not understand or explain, he never resorted to supernatural explanations.

One of the simple questions he proposed was: How does the sap rise up in the trees?

To answer this, Hales developed U-shaped glass monometers and measured liquid pressures at various distances on the tree. He found that there was a pressure difference between the root and the top that helped the sap to be "sucked up," as he put it.

In another study, he wished to find out how trees grow. To answer, he scratched horizontal marks on the stems of young trees at equal distances, and at monthly intervals, measured the distances between the marked intervals. He discovered that the middle segments had gotten farther compared to the other segments, and correctly concluded that most vertical growth in the trees occurred in their mid segments. He published these findings in 1727 in a volume titled *Vegetable Statisticks*.

So, how did tree sap and tree growth lead Hales to discover blood pressure?

One of Hales' sad duties in the church was to let sick and old dogs, sheep, stallions, mares, cows, and geese belonging to the church die by bleeding them. As the animals died, Hales studied them. He described his first experiment in his 1773 publication, *Statistical Essays: Containing Haemastatisticks*, which begins:

> "In December I caused a mare to be tied down on her back... having laid open the crural artery [the femoral artery near the thigh] about three inches from her belly, I inserted...a brass pipe whose bore was one fifth of an inch in diameter... [and connected to it another] glass tube, nearly the same diameter, which was nine feet: Then, untying the ligature on the artery, the blood rose in the tube eight feet three inches above the left ventricle of the heart..."

Stephen Hales Measuring Blood Pressure

After more such studies, he found that the *height of the blood column* changed as the heart contracted and relaxed—or systole and diastole. It also fluctuated when the animal struggled and breathed irregularly.

Although Hales' discoveries have now been ranked along with those of William Harvey, who described how blood circulated in the body, the importance of blood pressure for health and disease was not appreciated for more than a century later. In 1896, an Italian physician, Riva-Rocci, developed techniques to measure blood pressure in humans, and in 1905, Nikolai Korotkoff of Leningrad improved upon it.

Why did a country priest bother to study matters that were so worldly? In *Haemastatisticks* Hales tells us:

> *"The study of nature will ... [help us] search into and contemplate His works, in which the farther we go, the more we see the Signature of His Wisdom and Power, everything pleases and instructs us, because in everything we see a wise design...*

> *"...So complicated and curiously wrought a fabrick as an animal body is, the admirable and amazing texture of every part of which, declares its divine original..."*

Hale's religious path to understand the nature of God helped him to discover the signature of biology's wisdom and power.

A Matter of Life and Breath

In *Treatise on Man*, published a decade after his death, René Descartes (1596-1650), the scientist-philosopher, said that he believed that respiration was "extremely essential for the preservation of life." But he was skeptical about the importance of the heart and of the concept of circular motion of the blood described by William Harvey.

Despite Descartes' skepticism, other contemporary scientists continued to explore the physical world, making advances in physics and chemistry which helped research on biological topics. Two such examples concern pulmonary and cardiac physiology.

Robert Boyle (1627-1691) was an early member of a group called the "Invisible College," later to become the Royal Society of London. Boyle lived in an intensely rich intellectual atmosphere. Along with his pupil Robert Hooke (1635-1703), Boyle designed experiments to understand the nature of air animals breathe, and what would happen if the animals had no air to breathe at all.

Hooke constructed an elaborate piece of equipment, the "New Pneumatic Engine," with which he could create a vacuum chamber. Boyle and Hooke showed that when placed in the vacuum chamber, lighted candles were extinguished, and larks, sparrows, and mice died.

To further confirm their findings, Hooke volunteered to sit in a vacuum chamber while the air from it was being emptied. He sat until he developed earaches, and then ran out.

Based on their work in vacuum chambers, the duo proposed that the primary purpose of breathing was to bring fresh air into the body through inhalation, and to get rid of waste products through exhalation. But they were not sure how the lungs were able to exchange good and bad air.

John Mayow (1643-1679) was the third major scientist to contribute to understanding the nature of air we breathe. In spite of being one of the greatest scientists of the 17th century, Mayow received little credit for his work during his life and remained a controversial figure after his death.

Mayow joined Wadham College at the age of 17 and spent 16 years studying respiratory physiology in London and Oxford. In his book *TractusQuinque*, Mayow provides the first definitive explanation about the purpose of breathing.

In an experimental set up slightly different from that of Boyle and Hooke, Mayow placed a candle in a water-filled vessel with the wick of the candle a few inches above water level. Separately, he took an inverted glass jar, in the middle of which a horizontal rod had been inserted. He made a diaphragm-like platform from non-combustible material which could be hung from the glass rod in the jar. On that platform he placed a mouse; after lighting the candle in the water vessel, he placed the inverted jar (containing the mouse on the platform) over the candle, making sure there was plenty of space between the candle flame and the platform on which the mouse sat. He noted that after some time, the burning candle was extinguished, and yet a few minutes later, the mouse died. He then concluded that the candle was extinguished because the upper part of this setup had been deprived of "nitro-aerial particles." He had very vague notion of what those nitro-aerial particles were.

Moreover, he did not realize that the burning candle in the closed container had utilized oxygen and produced carbon dioxide, which had been dissolved in the water, such that the water level actually had risen up to fill the space lost by the burning of oxygen. By these experiments, he had come very close to discovering the existence of oxygen.

Based on the findings of his study, Mayow made a gigantic leap in speculation about the "nitro-aerial particles," implying just as those particles "entered" the candle, they would enter the blood after one takes a breath:

> "...on further consideration, I prefer the view that aerial particles enter the mass of the blood and they are deprived of their nitro-aerial particles...There is no denying the entrance of air into the blood [from] the lungs."

Initially, some contemporary scientists doubted Mayow's originality of research findings, but almost a century after his death, John Mayow was recognized as a preeminent physiologist, comparable in stature to Harvey, Boyle, and Sylvius.

Barcroft—the Man and his Mountain

When he turned 60, instead of contemplating a life of retirement, the British physiologist Sir Joseph Barcroft (1872-1947) started a new line of research—the study of the physiology of the fetus. The discoveries he made over the next 15 years became the foundation for all research related to the development of the human fetus.

Barcroft was born into a Quaker family and earned a BSc in London in 1891. An honor student all his life, he opted for a career in physiology research instead of medicine. He started as a lecturer at Cambridge's King's College in 1900, later became a professor, and was elected Fellow of the Royal Society for Improving Natural Knowledge in 1911.

Barcroft's field of research was high-altitude physiology. He tried to understand how humans and animals adapt to very high altitudes, known as acclimatization. He led his team of scientists on expeditions, built laboratories on tops of mountains, and studied the changes in breathing, blood pressure, and cardiac functions at high altitudes.

His team climbed and set up study stations on Tenerife and Monte Rose mountains in Europe, and the Andes in Peru. Barcroft invented a machine he called *Differential Blood Gas Manometer*, a forerunner of modern machines that measure levels of oxygen, carbon dioxide, and acidity in the blood. From his work, scientists learned how gases inhaled into the lungs are transported into the blood and how they are delivered at the tissue level. This became the foundation of modern heart-lung physiology. His 1914 monograph *Respiratory Functions of Blood* is a classic, which brought him world renown.

In 1932, Barcroft changed course and began studying the physiology of the fetus.

By then it was known that, when compared to the postnatal period, the fetus thrived in an atmosphere of very low oxygen concentration—less than one-half of extra-uterine oxygen levels in the blood. The mechanisms by which the fetus could survive in such conditions were unknown at that time.

Barcroft famously said that the fetus was living on "Mount Everest in-utero." Yet, he noted that "...one day, the call will come and the fetus will be born." Therefore, he wanted to explain how the fetus survived, grew, adapted during labor and delivery, and survived in the postnatal period.

Barcroft developed new animal experimental models and launched a comprehensive series of research studies. A brief list of topics he explored included: fetal blood volume; cardiovascular reflexes; cardiac output and its distribution; the physiology of the ductus arteriosus; fetal growth and utilization of glucose, fat, proteins, and amino acids; placental exchange of nutrients; properties of fetal and adult hemoglobin for carrying oxygen and carbon dioxide; and the physiology of fetal breathing.

During the process, he trained dozens of students from all over the world who, upon returning to their home countries, established major centers for research and training in human physiology.

In 1939, with Great Britain in WWII, Barcroft's research stopped abruptly. But he continued writing; he said, "...during the days of the bomb [ing], an hour or two each day to relax and write...information concerning prenatal life. [Because]...if the bomb came my way, the information for which it was worth, would [still] remain."

No bomb came his way. After the war, Barcroft published a monograph, *Researches on Pre-Natal Life, Volume I*, with plans for

Volume II on the fetal nervous system. He died from a heart attack on March 21, 1947, a few weeks after *Volume I* was published.

In 1954, the US Board on Geographic Names christened *Mt. Barcroft* for a 13,040-foot (3979 m) peak located at latitude 37.58220°N and longitude 118.2482°W in California's White Mountain Range.

Even though Dr. Barcroft never visited the United States, there exists the only mountain peak in the world named after a physiologist.

It Goes in Circles

Contrary to popular belief, William Harvey (1578-1657) did not discover circulation of the blood. He drastically *revised* the concepts of circulatory physiology and inaugurated the renaissance in physiology research methods.

Harvey's 1628 classic, *An Anatomical Treatise on the Motion of Heart and Blood in Animals*, remains a testimony to his genius. In addition to explaining the results of his experiments, Harvey also meticulously details how he did them—perhaps the first research manual to share experimental methods with uncanny precision. Because of this, even after nearly four centuries, we can see the logic behind his conclusions.

Prior to Harvey, the blood circulation was poorly understood. The concepts consisted of a set of confused assertions with little experimental proof. A typical student of physiology learned something like this about the circulation of blood and the functions of heart and lungs:

- Liver converts the energy from food into "vital blood," which flows through the veins to all parts of the body by way of an "ebb and flow" mechanism.

- The lungs receive this blood from the heart through the pulmonary veins, mainly for the lungs' own nourishment.

- The air we breathe in, or the "inspired air," flows from the lungs to the left side of the heart via the pulmonary veins, and mixes with blood by pouring out through "invisible holes" in the wall that separates the ventricles—much like pores in our skin for letting out perspiration.

- The arteries carry air mixed with blood and the heat generated by the heart's pumping action. (The term "artery" actually means "air I carry.")

- The brain converts blood into "animal spirit," which flows inside the nerves and helps the animals to walk, speak, use muscles, limbs, and so on.

- The heart is the center of emotions, which is why it beats faster when one is angry or in love.

- Because the heart is the center of emotions, the vocal cords, which express emotions through speech, are closer to the heart than to the brain.

- The inhaled air and the fanning motion of the lungs help cool off the heart and prevent it from becoming excessively warm.

What a mess.

Harvey's first clue to the true nature of circulation probably began when his teacher Fabricius taught that the cusps in the valves of veins always opened towards the heart, thus enabling blood to flow only *towards* the heart, and preventing its backward flow.

Harvey studied the motion of the heart in many different animals directly, or *in-situ*. From this he concluded that when the ventricles contracted, blood is ejected *simultaneously* into the pulmonary artery from the right ventricle, and to the aorta from the left heart chamber.

With this observation, Harvey concluded that the venous flow must always be *towards* the heart, and the arterial flow always *away* from the heart. If the contracting heart sent out blood only to the arteries, he theorized, the overall motion of the blood had to be one way, such that the directions of arterial and venous flows remain opposite to each other.

To find out how much blood was ejected from the heart chamber each minute, he removed the heart from a corpse and measured its volume—how much blood it can hold. He multiplied this number by the number of heart beats per minute, and discovered that each hour the left side of the heart ejected about 8,640 ounces—or about 4.4 liters each minute. This value is remarkably close to the value of cardiac output we now know.

Using the above computations Harvey surmised that it would be absurd to expect the liver to "pump" 8,640 ounces of new blood each hour and control the flow of so much blood through the veins. The logical conclusion from this had to be that it was the *same blood* that flowed in the veins and came back to the arteries.

So, the blood really "circulates"—it goes in circles, round and round!

In his book, Harvey explained his experiments and gave supporting data without any philosophical arguments.

To appreciate Harvey's genius, consider that the microscope had not been invented (thus he could not see capillaries), and oxygen, carbon dioxide, and nitrogen had not been discovered.

Harvey's contribution set in motion an entirely new direction of thinking and of asking questions: Why does the heart beat? How does blood go in circles? What is the function of breathing? What happens to the blood in the lungs that makes it change color from purple to red?

It took nearly two more centuries for scientists to find those answers.

Brother, Could You Spare a Kidney?

In Boston, on December 23, 1954, Dr. Joseph Murray (b1919) removed a kidney from a healthy man and transplanted it into the abdomen of the donor's identical twin, whose kidneys had failed. The operation was a technical marvel, and the transplanted kidney began working instantly. The donor remained healthy, which gave hope to thousands of future patients.

In spite of its success, the procedure raised ethical questions: What if the transplanted kidney had failed or the donor had died after removing one kidney?

The history of grafting has been tumultuous. Susruta (c 200 B.C.), an Indian surgeon, repaired amputated noses and torn earlobes by dissecting layers of patients' own skin carved out of the forehead or scalp. He would strive to maintain the circulation to the grafted flaps from the base of the grafted skin, such that the amputated defects were completely covered and the final results were spectacular, even by today's standards.

But the notion of organ transplantation belonged to the realm of mythology. Legend has it that Saints Damian and Cosmos, the doctor-brothers martyred in the 4th century, amputated the ulcerated leg of a Caucasian and replaced it with that of a dead Moor.

St. Damian and St. Cosmos Perform a Transplantation

In Thomas Mann's *The Transposed Heads*, the heroine obeys a goddess' order to resurrect two slain friends who had decapitated themselves over their love for the heroine. In her haste, however, she places the head of one friend onto the neck the other, and the eternal dilemma of mind over body unfolds. Ultimately, the mind was the winner.

Animal-to-human organ transplantation is told in the Indian

33

legend of Ganapati, the elephant-head god. He was the son of Parvati, wife of God Siva, created by her from the dirt on her skin to guard the doors while she took a bath. When Siva returned, he found the unknown child guarding the door refusing God Siva's entry. Outraged, Siva beheaded the boy, only to learn later that he was the son of Parvati—and hence his own.

To rectify his misdeed, Siva ordered his "ganas" (the gangs) to go south and bring the head of the first "person" sleeping with his head to the north. The first such creature happened to be an elephant, and Siva's servants brought its head back, which was transplanted on the trunk of Parvati's son, infusing life again—Ganapati, wisest of all gods and the lord of all of Siva's creatures was born. (Because the unfortunate elephant lost its head while sleeping with head to the north, the Hindu custom forbids one to sleep with one's head to the north.)

Organ transplantation in real life was far more complex than following a deity's command. Finding a donor was a major issue: Should it be human or beast? Would organs from dogs, pigs, or monkeys work just as well in humans? Could a human donor survive the sacrifice?

Alexis Carrel (1873-1944), a French surgeon and philosopher, provided an early breakthrough. After learning that the president of France had died from bleeding of the portal vein because no surgeon could suture it, Carrel, a surgery student in 1894, was so saddened that he determined to devote his life to experimental surgery. Within a decade of working on animals, Carrel had perfected a technique called the "purse-string method" to suture large, torn blood vessels without causing scars.

To repair torn blood vessels in animals, he grafted rubber tubes onto the vessels. Extending the idea of transplantation, he began removing whole organs, such as the thyroid gland, spleen, ovary, and kidneys, and transplanted them to animals from which the same organs had been removed. One of his experimental dogs lived for 30 months with a single transplanted kidney. He demonstrated such procedures were technically feasible.

Quite often, however, the transplanted organs were rejected by the host, which remained a vexing problem. The mechanism by which the host recognized the transplanted organ and destroyed it remained unclear until three scientists in pursuit of unrelated issues provided the answers.

In 1901, the Austrian scientist Karl Landsteiner (1868-1943) discovered human blood groups and established the uniqueness of red blood cell types. Later, other scientists showed that in cattle twins, the blood groups were always compatible.

Taking this lead, Sir MacFarlane Burnet (1899-1985), an Australian virologist, proposed the so-called "clonal selection theory," according to which a tolerance to a foreign tissue occurs from exposure to its proteins during early embryonic life. Sir Peter Medawar (1915-1987) of Great Britain used skin grafting experiments in mice and proved Burnet's theory correct.

When World War II veteran and Harvard Medical School graduate Joseph Murray began transplantation work, some of the puzzle concerning transplanted organ rejection had been solved. From research in animals, Murray first showed that transplanted kidneys worked well in the recipient animals.

By choosing an identical twin as donor for his historic 1954 kidney transplantation, Murray had hoped that the organ would be accepted—and he was right. After five more years of research, he also showed that kidneys from fraternal twin donors worked well if the immune system of the recipient was suppressed by total-body radiation.

By the early 1960s, a new class of drugs had been developed, notably by doctors George Hitchings (1905-1998) and Gertrude Elion (1918-1999) of the Burroughs Wellcome Laboratories, to treat various forms of cancers. Scientists were learning quickly that drugs used to suppress or cure cancer were also suppressing the entire immune system.

Using this knowledge, Murray made a pivotal discovery in 1962. He showed that after tissue matching, kidneys from *unrelated* donors too functioned well in the recipients if one used combinations of medications to suppress the immune system in the recipient—these medications were the same ones used in cancer patients.

The 1990 Nobel Prize in Physiology or Medicine was given to both Murray and E. Donnel Thomas (b 1920), who developed bone marrow transplantation. Other scientists who researched transplantation were also awarded the Nobel Prize: Carrel in 1912, Landsteiner in 1930, Burnet and Sir Peter Medawar in 1960, and George Hitchings and Gertrude Elion in 1967.

After decades of remarkable success of organ transplantation, an ugly side to this story has surfaced: the business of organ trafficking, the removal of organs without donors' consent and knowledge, as well as so-called transplantation tourism.

In 2001, *The Washington Post* reported that a Chinese doctor seeking asylum in the US confessed to being a part of a scheme involving removal of organs from patients without their consent or knowledge in China. More such reports surfaced, including claims that the removal of organs from Chinese death row prisoners prior to their execution had become a common practice. International uproar followed, and the Chinese authorities instituted new guidelines banning this practice.

But black market organ trafficking continues.

These days, with easy access to social media, one can market just about anything. Without strong international barriers and regulations, the end of trading body parts seems far away in the 21st century.

The Belly-button Mystery

No pit or dimple in the body has caused more mystery and confusion than the navel.

During much of the second millennium, artists faced a vexing dilemma: when painting Adam and Eve, should they or should they not depict the belly-button? Painting the navel would be blasphemous, but leaving the belly bare would be inaccurate, and even funny.

The problem was more than skin deep. If the artist chose to show navel dimples in Adam and Eve's paintings, the implication would be that Adam and Eve were connected to their respective mothers through the umbilical cords. This would contradict the Biblical story of God's creation of Adam, and Adam's creation of Eve. If the artists chose to leave the belly bare, they might even imply that unlike others, God's first humans were imperfect.

Artist's dilemma aside, the legends surrounding this body part are colorful.

In Hindu mythology, Lord Brahma, the creator of the universe, was born atop a fully bloomed lotus, the stem of which was rooted in Lord Vishnu's navel. The symbolism implied that all life forms are interconnected through the umbilical cords and are attached to their respective pasts in an eternal cycle.

In the Greek mythology pertaining to the navel, or the *omphalos*, Zeus wanted to know the location of the center of the earth. He released two eagles from opposite ends of the earth's perimeter, ordering them to fly at equal speed towards each other until they met. The eagles flew and met at Delphi, which was then declared the "center of the earth." The place was marked by a marble stone (*omphalos stone*) at the Temple of Apollo. It is a half-egg-shaped marble, guarded by golden eagles on each side. The omphalos stone appears on many ancient Greek coins and art objects. Eventually, the term "omphalos" came to mean the "center of all things," symbolizing the origin of life. Fittingly, the prefix "omphalos" is used to denote pathological conditions related to belly buttons (e.g., "omphalitis" means infection of the belly button).

Similar to its origin, the surgical conditions of the umbilicus, too, had been difficult to understand and treat. One such condition is omphalocele, in which the small intestine of the fetus migrates (as a hernia) into the naval cord. The resulting swelling can be as small as a walnut or as huge as a cantaloupe, depending upon the extent of gut hernia.

Another condition is gastroschisis, in which abdominal organs hang outside the abdomen through a defect in the abdominal wall near the umbilicus; the navel cord itself remains normal. Depending upon the size of the defect, the infant's stomach, liver, spleen, and intestine—or all—may lie outside. Both these conditions can be diagnosed by ultrasound conducted before birth, and surgery can cure most infants.

But, for centuries, omphalocele and gastroschisis were constant sources of confusion for the pathologist, who could not differentiate between them, and frustration for the surgeon, who did not know how to treat them.

With the exception of very small omphalocele, the mortality was uniformly high. Dr. William Hey from Leeds, England, first attempted to treat infants with umbilical hernia. He had been trained by Sir Ashley Cooper, a famous surgeon who had perfected the technique for surgically treating inguinal hernia in adults.

In 1803, Hey described three cases of "congenital umbilical

hernia." Two of the three infants died within a day of birth, despite an attempted "reduction" of the hernia—it is unclear how he attempted it.

For the third infant, Hey pushed the intestine inside the abdomen, and as his assistant pinched the base of the navel, Hey used "circular pieces of leather" on which he spread plaster in a concentric manner, then tied a linen bandage around the skin surrounding the navel. He noted that after the treatment the navel appeared like a perfect "cicatrix," or a scar. He declared that the infant had been cured.

Other surgeons, however, feared operating on infants. In the days prior to anesthetic agents, children who underwent surgery suffered a very high mortality rate. Surgery for newborn infants did not begin until the mid 20th century.

Even then, the evolution of children's surgery was far from smooth. General surgeons objected to the creation of surgical subspecialties. Dr. Edward Churchill, a surgeon from Harvard Medical School, remarked that his residents at Massachusetts General Hospital "... were quite proficient at operating on rabbits," and therefore there was no need for subspecialty in pediatric surgery.

It was Harvard Medical School, however, that founded America's first department of Pediatric Surgery in 1941 with William E. Ladd as its chair. Dr. Ladd and his pupils were the pioneers in neonatal and pediatric surgery and proved that operating on infants and children was vastly different than operating on rabbits.

In 1948, Robert E Gross (1905-1988), a pioneering pediatric surgeon at the Peter Bent Brigham Hospital in Boston, used mobilized skin flaps to cover a large omphalocele; 20 years later efforts at surgery for gastroschisis began in earnest, yielding regular success. Improved neonatal care along with intravenous nutrition support were greatly responsible for continued improvement in survival for babies with gastroschisis and omphalocele.

Back to the medieval artists' dilemma of painting navels for Adam and Eve: They used clever strategies to cover up the navel dimple. They reasoned that because of the "original sin," Adam and Eve would have felt ashamed of their nudity and would have covered their genitals with their hands. By artistically inserting twigs and leaves into the couple's hands, they could shield the troublesome navel dimple.

In fact, by hiding Adam and Eve's navels, the artists succeeded in preserving their own hides.

In 1969 a south Indian scholar writing in the Kannada language famously titled his first collection of poems *No Lotus in the Navel*,[1] mischievously implying that as a spontaneously-born poet he had no lineage and thus was peerless. The collection was critically acclaimed.

Things indeed have changed.

Figure 1: *Adam and Eve in Paradise* (tempera on panel),
Scorel, Jan van (1495-1562) License and courtesy: Johnny van
Haeften Gallery,
London, UK. Bridgeman Images.

Figure 2: Lord Brahma on the lotus connected to Lord Vishnu through the lotus stem attached at the umbilicus. Courtesy, Mr. Kyle Tortora. **www.lotussculpture.com.** Bronze; total height including the base, 9″; Base width & depth, 15″X 6″; weight 18 pounds.

Figure 3; Omphalos stone, Delphi. Thought to be a (Roman?) replica of the original sacred stone which mysteriously disappeared and was displayed outside the temple.
Photo by **Юкатан** 2009. Via **Wikimedia**

References:

1. A. K. Ramanujan. *No Lotus in the Navel* in, Raju T.N.K. and Daniels-Ramanujan, M. (Translators); *Poems and a Novella by A. K. Ramanujan*, Oxford University Press, New Delhi, 2006. Pages 3-58.

Acknowledgement: This article was previously published in *Hektoen International,* **Volume 13, Issue 2 – Spring 2021**. I thank the editorial team for the permission to reproduce it. I also thank those entities that gave copyright permission to reproduce the figures in this article.

Double Helix and Beyond

The most exciting, but extremely boring encyclopedia made its appearance in the final year of the 20th century.

This 425-page volume contained 36 lines per page and 65 letters per line. Four thousand such volumes together make up the encyclopedia, in which only four letters are used—A, T, G, and C. In this first offering of the Human Genome Project, a monotonous outpouring of chemical formulae described about four billion base pairs in some 100,000 human genes. Not a *New York Times* bestseller—yet, this encyclopedia of the human genetic code was the most eagerly anticipated achievement of 20th-century medicine.

The successful culmination of the Human Genome Project can trace its origin to the biochemical revolution that began in the late 19th century and ended with the discovery of the double helix structure of the DNA molecule in 1953. The spectacular conclusion of the Project owes its debt to technological and conceptual advances in molecular biology and computer sciences during the post-double helix era.

Children inherit their parents' features—this had been known since antiquity. But how such inheritance occurs remained a mystery.

Ancient Greeks thought that parents' traits were transferred via the blood into the next generation—thus the term "bloodline."

The modern era of genetics began with the discoveries of two

19th-century gentlemen: Charles Darwin (1809-1882) and Gregor Mendel (1822-1884). Charles Darwin's Theory of Evolution (1859) and Mendel's work on transmission of hereditary features in garden peas (1865) were the foundations upon which later discoveries would be built.

Mendel, an Augustinian monk who taught natural science to high school students, was interested in mathematics, meteorology, and evolutionary theories. He wondered how traits passed on from parents to the progeny in the plant kingdom. Over a seven-year period, he cross-bred thousands of peas of different colors and sizes and derived the basic laws of heredity.

Mendel concluded that traits were inherited in precise, predictable mathematical ratios. He wrote that the "heredity elements" were transmitted intact, such that the progeny manifested the "dominant" feature, suppressing the weaker trait. Only in the absence of a dominant feature did the weaker trait manifest itself in the progeny.

Despite such elegant results, Mendel's work was ignored and forgotten, while Darwin's theory of evolution was misunderstood and remained controversial.

In 1900, the discovery of chromosomes led to the concept that those thread-like elements may bear Mendel's "heredity elements," which were later called "genes."

One day in 1910 in his "fly room," Thomas Hunt Morgan (1866-1945), an American biologist who studied chromosomes in fruit flies (*Drosophila melanogaster*), noted a male fly with white eyes rather than the usual red. Upon breeding this fly with a red-eyed female, all of the first-generation offspring born had red eyes. Inter-breeding that first generation of offspring led to a second generation with both red and white eyes in a 3:1 ratio; all white-eyed flies were males.

Morgan had discovered sex-linked inheritance and shown the link between genes, traits, and chromosomes.

At that time, most scientists thought that genes were proteins. Oswald T. Avery (1877-1955), a Canadian-born American scientist, altered this view in 1944. When he mixed heat-killed, disease-producing pneumococcal bacteria with living, but benign, strains of pneumococci, the latter "learned" the art of causing the disease from the genes of the dead bacteria. Since heat destroyed proteins, Avery concluded that genes were *not* proteins, but perhaps linked to heat-resistant nucleic acids, or deoxy nucleic acid (DNA). The Swiss biochemist Johann Friedrich Miescher (1844-1895) had discovered "acids" in the nuclei of cells in 1868—thus Avery concluded that they were "nucleic" acids.

By 1930, others had shown that the nucleic acid molecules contained equal amounts of the base units adenine, guanine, thymine, and cytosine (A, G, T, C). Until Avery's work, no one had seriously considered that so "ordinary" a chemical (polymer) as DNA could be the substance involved in transmitting hereditary traits.

Avery's discovery that DNA may be the source of genes unleashed a race to decipher the physical structure of the DNA molecule. By the early 1950s, using X-ray crystallography, the British physicists Rosalind Franklin (1920-1958) and Maurice Wilkins (1916-2004) photographed the DNA molecule. BBut it was Francis Crick (1916-2004) and James Watson (b 1928) who brilliantly modeled the double-helical structure of DNA in 1953. Watson had seen Franklin's DNA photograph at least once before.

Watson and Crick's three-page report in *Nature* ends with one of the most profoundly understated remarks in medical history: "It has not escaped our notice that the specific pairing we have postulated [for DNA]... suggests a possible copying mechanism for the genetic material."

Soon after the discovery of the double-helix structure of the DNA, questions never before considered began to confront the scientific world: Why a helical structure? Why are the building blocks so monotonously repetitive, yet unpredictable? What is the ultimate chemical structure of the gene? How exactly is the genetic message encoded and how is it replicated? Can genetic mapping be used to predict, prevent, and treat diseases?

By the mid-1960s, most of those questions were answered. Watson, Crick, and Wilkins won the 1962 Nobel Prize for their DNA work. Over the next 15 years, more than a dozen Nobel Prizes followed for DNA-related discoveries. Some of these achievements were: the discovery of the role of RNA in protein synthesis; discovering and using scores of molecular scissors—enzymes—to cut and trim long DNA segments and to stitch and repair them; and the ability to produce "designer" genes.

By the late 1960s, the complete base pair sequence of a gene had been deciphered. These advances have helped to identify chemical defects for hundreds of common, and thousands of rare, single-gene defects. Genetic markers are available for assessing the risk of developing multifactorial diseases, such as breast and cervical cancers, Alzheimer's disease, diabetes, hypertension, and heart disease. Gene therapy is around the corner.

With advances come dilemmas. If we can detect a disease early, should we apply the test to one and all? Do we stigmatize the carriers of such genes? Who assures confidentiality? Is it ethical to commercialize and sell genetic data?

Even though answers to these questions are critical, consider this: Until 1905, even the term "genetics" did not exist, and four years later, the Danish botanist Wilhelm Johannsen (1857-1927) coined the term "gene"; until the mid 20th century, the acronym DNA signified only a lowly chemical molecule with no known functions; and until 1955, we did not know the correct number of human chromosomes.

Yet, by 1990 the Human Genome Project was conceived and launched, and by 2003 it was completed.

A giant leap for mankind.

Immunity—Anti-immunity

By the end of the 19th century, there was great optimism that all diseases could be cured by a newly discovered property of the body—the immune system. Some scientists tried to develop vaccines and sera against all bacterial agents and inject them into bodies for developing permanent immunity.

People quickly learned that the nature of immunity was more mysterious. Although several anti-sera were developed to kill bacteria, scientists found that some patients upon receiving anti-sera died suddenly from violent reactions. The immune system that normally protects the body could also kill. This ugly side of the body's immune reaction remained a mystery until Charles Richet (1850-1935), a French immunologist, discovered "anaphylaxis," a term he coined which means "contrary to protection."

Richet was fishing on a cruise over the balmy waters of the equatorial sea in a yacht with Prince Albert of Monaco. They caught a Portuguese man-of-war, a sea anemone that has long tentacles with which it injects poison upon contact with its victim. The prince asked Richet to investigate the poison and develop an antiserum. They caught several of these creatures and brought them back to Paris.

Richet later extracted poison from the Portuguese man-of-war. He first injected minute quantities of the poison into laboratory dogs and showed that the dogs tolerated it well. But after additional injections, the same dogs went into unexpected violent reactions. Many suddenly died of shock.

Richet called this phenomenon "anaphylaxis" and said it to be an "inverse of immunity"—an extreme form of allergic reaction brought about by the body's immune system. He reasoned that this was perhaps the body's misguided way of protecting against an invading foreign agent.

The phenomenon of anaphylaxis was one of the fundamental discoveries in immunology. Charles Richet was awarded the 1913 Nobel Prize in Physiology or Medicine.

God in His Details

The study of anatomy involves the examination of organs and their relationship with other organs and with the rest of the body. But anatomy also requires the examination of the organ's basic structure as well as an understanding of its most fundamental unit.

The invention of the microscope helped answer some of the riddles related to the minute structure of organs. With the help of electron microscopy today, we can magnify the size of a period at the end of this sentence to appear bigger than a basketball.

But the microscope, even in its most rudimentary form, was not invented until the early 17th century. An Italian professor, Marcello Malpighi (1628-1694), started using the newly invented gizmo to study the lungs taken from animals. He prepared the microscopic sections by first flushing the lungs with water through an artery attached to the lungs (which helped force the blood out), inflating the lungs, then drying them before placing the specimen under the microscope.

The professor found so many details in the architecture of the lungs never seen before. He described what he saw in vivid details:

> "...when the [lungs] were cut inwardly, it reveals white mass of vesicles to all eyes...formed from the end of the trachea forming flask-shaped cavities."

Malpighi argued that the basic units of the lungs must be those tiny little sacs, not unlike individual grapes in a bunch. Later, these sacs came to be known as the alveoli.

Until Malpighi's description, what happens to the air we inhale was unclear. Anatomists had theorized that the inhaled air went directly from the trachea (through invisible holes) into small veins near the lungs, and from there to the heart.

Malpighi also showed that in other parts of the body, too, there were small blood vessels, the capillaries that connected veins and arteries. This discovery provided the finishing touches to William Harvey's proposition that blood flows in circles within an animal.

Dancing Bees and Baby Ducks

Two thousand years ago Aristotle asked: "How is the mind attached to the body?"

Three biologists in the 20th century working independently asked the same question with a slight twist: How do animals communicate with each other? How does one brain "talk" to the other?

For the Austrian biologist Karl von Frisch (1886-1982), the search began with a simple issue. At the turn of the 20th century, scientists believed that bees were colorblind. If that were true, wondered von Frisch, why did the flowers need to have vivid colors?

In a series of experiments in 1912, he placed sugar water on papers of different colors, and learned that bees can be "trained" to recognize colors and find the right kind of food. This proved that indeed the bees were not colorblind.

Over the next 15 years, von Frisch discovered that bees communicated with each other about their location and the nature of the forage with uncanny precision—the language of the dancing bees was beginning to be deciphered.

Another Austrian biologist, Konrad Lorenz (1903-1989), loved animals all his life. In the 1920s and '30s he studied the social behavior of birds, ducks, and their young ones. Using methods that are now classic, Lorenz showed that after the egg is hatched, the emerging ducklings and goslings followed any moving object, including Lorenz himself as he swam in a pond. He called this process "imprinting" by which the offspring adopted any object nearby as their mother. This was basic to understanding the phenomenon called "programmed learning," which explained how the adult brain passes on survival skills to the offspring's developing brain.

In a career spanning more than five decades, the Dutch scientist Nikolaas Tinbergen (1907-1988) studied animals ranging from digger wasps to sled dogs, from red-necked birds to large-winged butterflies, and from giant caterpillars to autistic children. His laboratory was any country that had wilderness: be it Greenland or the Netherlands, Germany or Great Britain, Tinbergen went there to study animals in their natural habitats. He developed revolutionary methods to study animal behavior. Tinbergen's 32-page research manuscript was the shortest thesis ever awarded a PhD at Leiden. A modest and soft-spoken man, it seems he felt embarrassed to have won the Nobel Prize, for he thought he was unworthy of such a high honor. In his acceptance speech, he stressed the importance of maintaining the art of "watching and wondering." He pleaded for more collaboration among various disciplines in the study of body and mind.

Together, Tinbergen and Lorenz thus founded the new discipline of neuroethology—a branch of neuroscience where one studies the neurobiology of animal behavior. Along with von Frisch, they shared the 1973 Nobel Prize in medicine for discoveries concerning animal behavior.

At the beginning of the 20th century, scientists were just beginning to learn how individual neurons "communicated" with other neurons. By the end of the century, the probe had become broader: attempting to learn the brain's language for communication with other brains.

We have just begun to find out some answers to Aristotle's question.

Reconstructing memories and history in *One Hundred Years of Solitude* by Gabriel García Márquez.

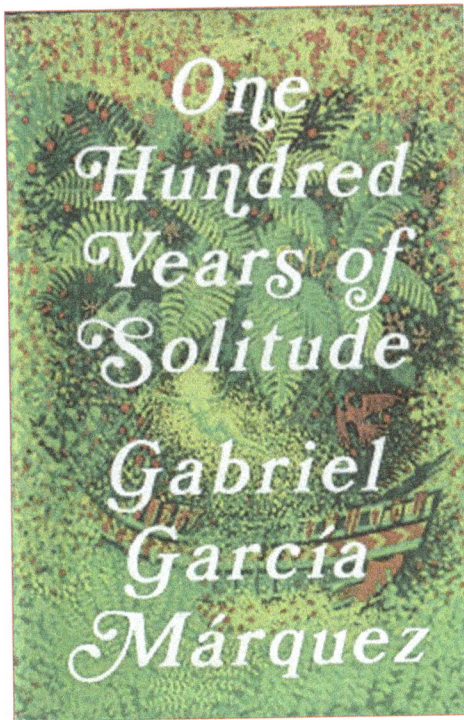

Figure 1: Gabriel García Márquez, *One Hundred Years of Solitude*, **1967**. Courtesy of Harper & Row.

"Many years later, as he faced the firing squad, Colonel Aureliano Buendía was to remember that distant afternoon when his father took him to discover ice."

In the opening sentence of his extraordinary masterpiece, Gabriel García Márquez distilled the recurring themes of *One Hundred Years of Solitude*[1]: the absurdity of death, the restorative power of memory, and the amazement of a discovery rolled into the eternal cycle of time. The omnipotent narrator begins in the present, telling us that "many years later" a condemned man would be facing a firing squad and remembers a seemingly trivial incidence: his discovery as a boy "that distant afternoon," as he touched a large cube of ice, which his father would later say was "the great invention of our time."

In this 417-page magnum opus [Fig 1], García Márquez weaves the stories of people's lives in a Colombian village, Macondo, in a prose style now famous as magical realism. The main story chronicles six generations of the Buendías, dense as the tropical forests of Latin America and spanning more than one hundred years. Passionate love, ruthless betrayals, unreal magical events, brutal insurrections, and mindless battles amidst revolutionary fervors—are these real or fiction? One wonders.

A single episode[1] that occurs early in the story invites a focus on contemporary sociopolitical significance: the struggles of the protagonist against the dual "plagues" of insomnia and amnesia, and the consequences of living in an alternate reality.

José Arcadio Buendía and his wife Úrsula offer shelter to Guaijiro Indian siblings Visitación and her brother. Both had fled a region stricken with the plague of amnesia. One day, Visitación warns the Buendías that her fatalistic heart has told her that the epidemic would follow her to the farthest corners of the earth. She, along with all in that household, would be affected by the plague. No one believes her.

A few weeks later, Jóse Arcadio finds himself unable to fall asleep. Yet, he feels no fatigue the next day. Soon all in that household become insomniacs, but they too do not worry since they never feel tired. When Úrsula sells her homemade candies in the village, the sickness spreads through those sweets and the whole village starts suffering from insomnia. Villagers spend days and nights doing nothing or playing silly games. Along the way, they forget how to dream. But being good citizens, they wish to prevent the sickness from spreading beyond their village. So, they place goat bells at the entrance to the village and ask visitors to ring them as they pass through Macondo to announce that they are uninfected outsiders.

Figure 2: A memory machine model (BSB Cod.icon 242, 40: 18r 0040) by Johannes, de Fontana from the manuscript *Bellicorum instrumentorum liber cum figuris*. Courtesy: Bayerische Staatsbibliothek, Munich, Germany; the collection in the public domain.[5]

Matters get worse when José Arcadio and his son Aureliano begin forgetting the names of familiar objects. To prevent loss of memory, they affix labels with the names of the objects: *table, chair, clock, wall, bed, pan.* But this is not enough and the loss of memory continues. They realize that soon they may forget what to do with those objects. So, they make more descriptive labels: *This is the cow. She needs to be milked every morning . . . the milk should be boiled to be mixed with coffee.* In this manner, they continue to live as the reality of life begins slipping away. In no time, however, the plague of amnesia affects all villagers. They, too, begin writing down the names of objects and their personal feelings, hoping not to see a day when they forget the value of written words and letters.

Defeated by all attempts to preserve memory, José Arcadio Buendía takes a dramatic new approach—building a memory machine, a spinning dictionary that could be operated by using a lever. With that machine, within a few hours the words and notions most necessary for life would pass in front of their eyes. He builds such a machine, writing almost fourteen thousand entries. Using it, every morning one could review the totality of knowledge acquired during a lifetime.

It was about that time that an elderly gypsy appeared at Buendía's house. He was well-known in Macondo, but thinking he had died, the villagers had forgotten him. The gypsy quickly saw that his old friend José was suffering from amnesia. He pulled out a small bottle containing a liquid of "gentle color" and gave it to José to take a sip. A light goes on in his memory—he is no longer amnesic. In no time, the remedy is passed on to the entire village. They all joyfully celebrate the restoration of their memories.

The story provides a perspective about the human cost of insomnia and amnesia. It also offers implied perspectives on the socio-political consequences of these formidable "plagues."

The people of Macondo did not feel exhausted despite being awake for days and weeks. Insomnia enabled them to finish the chores of daily living, but suddenly they found nothing else to do. Then they tried all kinds of tricks to fall asleep until the point of exhaustion—this was not from fatigue but because of their nostalgia for dreams. The inability to dream and enrich their lives was the heaviest price of insomnia.

But the price of amnesia was even higher. Individually, amnesia—the forgetting of self, surroundings, and childhood—became unbearable. The tragedy of personal memory loss was compounded when amnesia became an epidemic leading to the collective loss of memory, thereby forgetting history.

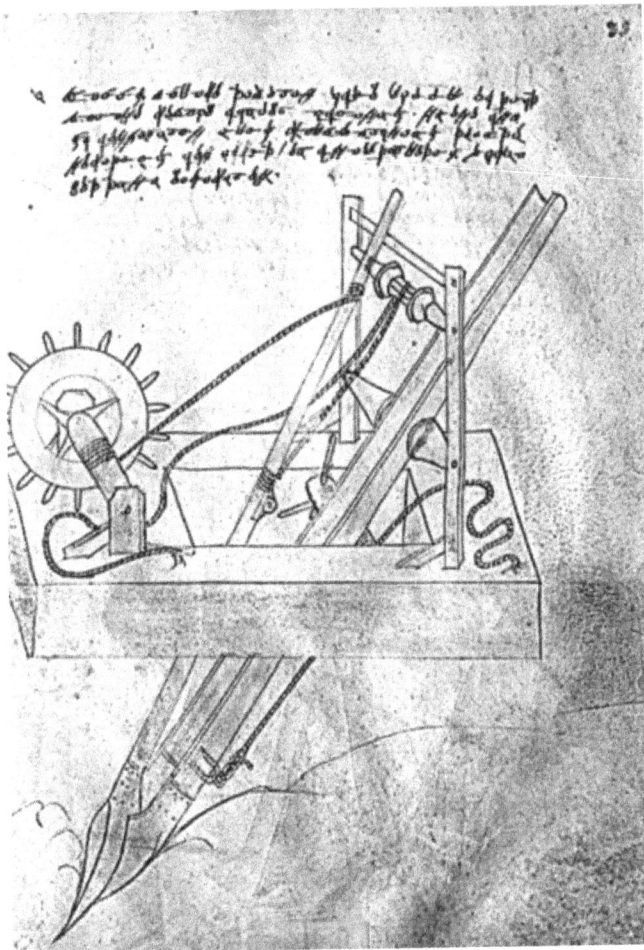

Figure 3: A memory machine model (BSB Cod.icon 242, 70: 33r 0070) by Johannes, de Fontana from the manuscript *Bellicorum instrumentorum liber cum figuris*. Courtesy: Bayerische Staatsbibliothek, Munich, Germany; the collection in the public domain.[5]

Memory and history are two sides of the same coin. Loss of historical memory and living in a world of alternate reality remind us of the oft-quoted George Santayana aphorism, which in its original form reads: "Those who cannot remember the past are condemned to repeat it."

The dangers of societal amnesia include the perpetuation of false beliefs and a failure to avoid future tragedies. There are many examples of historical amnesia, but a few contemporary examples highlight the dangerous consequences of this malady: minimizing or denying the horrific tragedy of slavery and racism in the United States, the denials of the Holocaust, forgetting the devastation caused by infectious diseases[2] such as smallpox, which killed more people in the twentieth century alone than all the combined wars and battles of that century, and the contemporary activism against all forms of immunizations and vaccinations.

The last component of the story with a medical perspective is the method the protagonist adapts to overcome the plague of amnesia. When all else fails, José Arcadio Buendía constructs a memory machine, a classic example of a mnemotechnic. Neuropsychologists have identified this phenomenon as synesthesia.[3,4] This is a process of removing boundaries between two sensory pathways, enabling one of them to trigger the other sensation. Examples of synesthesia include artists seeing colors as they listen to music, the sensations of the smell and taste of food upon seeing cooking shows on TV and visualizing the meaning of spoken words. José Arcadío's memory machine, a sort of revolving dictionary, is based on synesthesia. In addition, his attempts to write down names and meanings is an example of the best process to preserve historical memories. García Márquez likely modeled the concept of the memory machine, a forerunner of modern computers, by adapting the inventions of the brilliant fifteenth-century Italian engineer Giovanni Fontana[5] [Figures 2 and 3].

Did García Márquez intend for the readers of his stories to interpret them as described above? We can get a feeling for what he meant from his 1982 Nobel Prize speech titled *The Solitude of Latin America*.[6] García Márquez says that the entire Latin American reality is far stranger than fiction because despite the independence they obtained from the Spaniards, it did not put the continent "beyond the reach of madness." Nothing more needs to be added.

Just as the eldest son of José Arcadio Buendía and Úrsula did, their second son, José Arcadio, also faces a firing squad because of his revolutionary activities. As the smoking mouths of the rifles are aimed at him, he thinks of his pregnant wife Rebeca and says to himself: "Oh, God damn it! I forgot to say that if it is a girl, they should name her Remedios."

When the blood from the first bullet trickles down his thighs, he shouts, "Bastards!" as if to say, today I, tomorrow you.

References

2. García Márquez, G. *One Hundred Years of Solitude*. New York, NY: Harper & Row; 1967.
3. Hostetter MK. What we don't see. *N Engl J Med.* 2012;366(14):1328-1334.
4. Bolzoni L. The Play of Images. The Art of Memory from its Origins to the Seventeenth Century. In: Corsi P, ed. *The Enchanted Loom: Chapters in the History of Neuroscience* Vol Part 1. Oxford: Oxford University Press; 1991:16-165.
5. Luria A. *The Mind of Mnemonist: A Little Book about a Vast Memory*. Cambridge, MASS: Harvard University Press; 1987.
6. Fontana G. *Bellicorum instrumentorum liber cum figuris*. Bayerische Staatsbibliothek. https://iiif.biblissima.fr/collections/manifest/3b360000d703cf66516dc5d4eb9cdf1aa7c713e8?tify={%22panX%22:0.588,%22panY%22:0.687,%22view%22:%22thumbnails%22,%22zoom%22:0.355}. Published 15th century. Accessed September 12, 2021.
7. García Márquez, G. *The Solitude of Latin America*. The Nobel Foundation. The Nobel Speech Web site. https://

www.nobelprize.org/prizes/literature/1982/marquez/
lecture/. Published 1982. Accessed September 12, 2021.

Acknowledgement: This article was previously published in *Hektoen International,* **Volume 14, Issue 1 – Winter 2022**. I thank the editorial team for the permission to reproduce it. I also thank those entities that gave copyright permission to reproduce the figures in this article.

The Importance of Having a Brain

The World Health Organization declared the 1990s as the "Decade of the Brain." It has taken mankind several thousand years to understand the brain's importance.

The first written reference to the brain can be seen in the 3,500-year-old Edwin Smith Surgical Papyrus, in which a doctor describes various types of brain injuries and proposes treatments. He declares that he *will treat* some injuries because the patient will recover, others he says he *may treat* because the patient *may* recover, whereas the third group he *will not treat* because the patient will *not* recover. Perhaps this is the first identifiable reference to the "do not resuscitate" concept of the present day.

But the brain was not always considered important.

Even Aristotle, a great thinker, anatomist, embryologist, taxonomist, and scientist, blundered about the role of the brain in living creatures.

He argued that because of its rich blood supply and a "central location," the heart was the seat of the human soul and the center for all sensations. In contrast, the brain's relative pallor and peripheral location suggested to him its inferior status to the heart; hence since antiquity, the brain was deemed "cold" and relatively less important than the heart.

Why did Aristotle make such a profound mistake?

Aristotle was very "logical." Like many other logical people, he tried to fit his observations into his preconceived hypotheses. Here were his observations and "logical" interpretations.

The heart responds to emotions by beating vigorously, while the brain does not; therefore, the brain has no "emotions."

All animals have a heart, but only vertebrates and cephalopods have a brain. Yet, even "brainless animals," such as mosquitoes and other creatures, have sensation. Therefore, a brain is not required to feel sensations.

- The heart has lots of blood in and around it, and blood is essential for sensation; therefore, the heart must perceive sensation. The brain is pale with little or no blood as seen at autopsy. Therefore, it must not be required for sensory input.

- The heart is warm, suggesting a higher life form, while the brain is cold, and hence must be inferior to the heart.

- The blood vessels connect the heart with all organs, including all sensory apparatus, whereas the brain has no such direct connection with all other organs.

- Since it is the first organ to begin beating and the last to die, the heart must be essential for survival. It is in the center of the body, an appropriate location for its controlling function. The brain is situated at the periphery, obviously because it has no controlling role.

- The heart beats fast when pain is inflicted, whereas the brain, being insensitive, does not respond at all to pain—not even when it is pierced or cut. Since the brain cannot "feel" any pain, how can it sense pain?

- The eyes are located near the brain because the brain

is fluid and cold, as are the visual functions, which are "watery." Having eyes in front helps us to see and move along a straight line in front of us.

- The ears are located on either side of the brain because we can hear from all sides. It does not mean that the brain is needed for hearing; there are many animals with no ears at all, yet they can hear well.

For centuries, scientists wondered about Aristotle's errors; even Galen, one of the most famous doctors from the 2nd century CE said that he "blushed to quote" Aristotle on this topic.

But Aristotle's influence was so great that the concept of heart-centered physiology survived with little challenge for 15 centuries—from Ancient Greece through the Renaissance. Consider René Descartes' famous remark: "I think, therefore I am." He seems to have meant that "thinking" resided outside the realm of the body. With contemporary knowledge of brain and its function, perhaps he might have said, "I can think, *because* I have a brain" or "I am (or I have a brain), therefore I can think."

Shakespeare expresses the brain-heart dilemma in *The Merchant of Venice*:

> *"Tell me where's fancy bred?*
> *Or in the heart, or in the head?*
> *How begot, how nourished?*
> *Reply, reply."*

Despite the prevailing confusion about the brain and its function, it is remarkable that Franz Joseph Gall (1758 -1828), a brilliant anatomist, proposed the theory of cerebral localization by developing "phrenology," and perhaps inaugurated the modern era of neurophysiology.

In the 17th century, people believed that one's physical features and body composition predicted one's personality. Gall later said that he was impressed by the visual memory of one of his elementary school classmates, who had "big, bull's eyes." Because of this, Gall became convinced that facial and external features could predict personality traits.

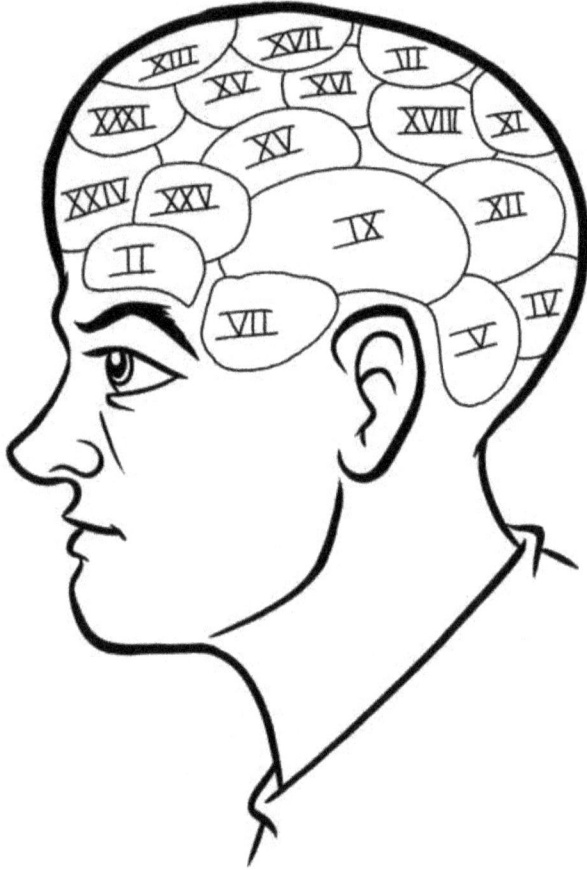

Gall's Concept of Brain Regions

Gall studied more than 300 skulls from people with a wide range of personalities and social positions, from thugs and criminals to brilliant teachers and professors. He concluded that by touching and feeling the bumps and depressions on his subjects' heads he could "read" their brains and understand such traits as destructiveness, veneration, and inquisitiveness.

Gall became extremely rich by "reading heads." His neurology practice had a specific brand name, the practice of "phrenology," which thrived until the mid-19th century. Queen Victoria had her children's heads read by phrenologists, and the famous writer George Eliot had *her* hair shaved to enable Gall to read her head.

However, for no clear reasons phrenology began to fade by the mid-19th century, only to be replaced by the theories concerning "cerebral localization," in which specific cerebral regions were considered to be controlling specific body parts.

In 1861, Paul Broca (1824-1880), a French neurologist, described a dying patient with hemiplegia (paralysis of one side of the body). The patient was unable to speak despite retaining motor powers needed for producing speech. When the patient died, Broca managed to obtain an autopsy, and found that the patient had a lesion (probably a tumor or a stroke) in the left anterior-inferior region of the brain. He concluded that "we speak with our left hemisphere." This region controlling the production and comprehension of speech is now called Broca's Area.

It would take a few more decades until Camillo Golgi (1852-1934) and Santiago Ramón y Cajal (1852-1934) would ignite the imagination of the scientific world by their independent research on how the brain worked.

In 1873, Golgi developed a technique to stain and prepare microscopic sections of the brain, called silver-impregnation technique. Later Cajal improved the method, and both these scientists independently described the microscopic architecture of the human brain. KorbinianBrodmann (1868-1918) and others refined the types of brain cells found in specific brain regions; thus by the end of the 19th century, the *neuronal doctrine* of the brain had been proposed.

According to this theory, the fundamental unit of the brain was a single cell—the neuron, while the neuroglia, blood vessels, and meninges were the supporting cast. In the 20th century, neuroscience then progressed with unprecedented speed and sophistication. Charles

Sherrington (1857-1952), the "philosopher of science," described gaps at neuronal junctions and in 1897 coined the term "synapse." During the next 50 years, studies on nerve conduction, spinal and central reflexes, and the biochemistry of neurotransmitters clarified many basic aspects of brain functions.

But what did the ancient Egyptians think about the importance of the brain? We cannot be sure—but for all their sophistication, it appears that they were ambivalent about the brain.

For instance, in the tombs for their pharaohs, they systematically preserved the heart, liver, spleen, kidney, and other organs in containers that were shaped like the organs being preserved, so that the departed pharaoh could use them during his afterlife.

However, the embalmers did not preserve the brain, which they meticulously removed piece by piece by inserting long needles through the roof of the nostrils.

Why did they not preserve the brain? Perhaps they felt that a brain was not necessary for the emperor's final journey.

It is reassuring that even then brainless heads-of-states were not uncommon.

No Humbug

Before the development of surgical anesthesia, surgery was a brutal craft.

Whether he was pulling an infected tooth or amputating a gangrenous leg, the surgeon had to wield his knife swiftly, finishing the job while the powerful hands of his medical students and assistants forcibly restrained the screaming patient. A surgeon's compassion was measured by the speed of his knife. Thus, only the most desperate patients sought surgical relief.

This changed completely after the work, discoveries, and the tangled lives of Horace Wells (1814-1848) and William Morton (1819-1868), two New England doctors.

On December 10, 1844, Wells bought a ticket for a show at the Union Hall in Harford, Connecticut. Billed as a "genteel demonstration of the effects of the laughing gas," the show promised great entertainment, featuring nitrous oxide, which had been discovered in the 18th century.

Wells sat on a back bench, relaxed, and looked forward to having a good time. He was not disappointed.

Gardner Colton, the owner of the show, had invented a contraption to administer the nitrous oxide. After an impressive speech about the powers of laughing gas, he invited volunteers from among the spectators for the demonstration.

Samuel Cooley, a drugstore clerk, agreed. Colton administered the gas; after inhaling it for a few minutes, Cooley made a complete fool of himself. His gait became unstable and speech slurred. Making absurd jokes, he went into a laughing fit and fell off the podium. Even then, he continued his non-stop giggling and telling of jokes. Spectators loved the demonstration, watching a perfectly normal man transform and make a perfect ass of himself.

Wells enjoyed the show too, but began to see something else: Cooley, who had fallen and sustained a large bump on his leg, seemed not to be bothered by pain. He acted as if there was nothing wrong with his body.

As a dentist, Wells knew how his patients suffered from dental pain while their infected teeth were being pulled out with no anesthetic agent. Wells wondered whether laughing gas could eliminate such pain among his patients.

After the show, Wells asked Colton (who also traveled with P. T. Barnum and his circus) to participate in an experiment using laughing gas for a painless tooth extraction. After some hesitation, Colton agreed.

But on whom should they try the gas? How should they pull the tooth if the patient "laughed" all the time? They feared that they may need to keep the patient still with a hefty dose of alcohol. But what if the patient died from excess alcohol?

Since there were no simple answers, there was only one choice—Wells himself had to be the guinea pig for dental extraction.

Dr. John Riggs, a fellow dentist, agreed to extract Wells' tooth. After Colton administered a generous dose of nitrous oxide, Wells' tooth was removed from the jawbone without a hint of pain. Arguably, this was the first tooth extraction in history after which both the surgeon and the patient were happy. It was also the first proof of the anesthetic property of the laughing gas.

Wells and Riggs treated more dental patients using laughing gas and were delighted with their success. But they needed to share their discovery with the world.

So in 1845, they asked Dr. John Collins Warren (1778-1858),

a famous surgeon at Harvard Medical School, for a chance to demonstrate their discovery of painless tooth extraction in front of medical students. After some reluctance, Warren agreed to the demonstration.

A volunteer "patient" had to be chosen: an unwilling medical student.

Colton administered laughing gas and Wells removed the student's tooth. During pulling, the student flinched and moaned, indicating lots of pain.

The students shouted, "Humbug!"

The tooth had come out, but the patient seemed to be in pain. Feeling dejected, Wells left the operating theater muttering, "But it works!"

But the humiliation of being taunted in front of a large audience haunted Wells' sensitive soul the rest of his life.

Enter William Morton:

A former dental apprentice of Wells, Morton was now a medical student at Harvard. He was among the spectators watching Wells' botched demonstration. Morton immediately guessed what had gone wrong—the laughing gas had indeed worked, but probably the dose was low and the scared volunteer had overreacted with his display of pain and discomfort.

Morton went to work in secrecy. He capitalized on Wells' discovery of the anesthetic property of chemical agents, and decided to develop his own contraptions to administer the anesthesia. He partnered with an engineer and developed a jar with two neck-like stems on either side. Through one opening, the patient could inhale an anesthetic agent, while through the other the doctor could add the agent to the jar.

Morton went back to Harvard Medical School, where he sought and received permission for a repeat demonstration of the effects of his "secret" anesthetic agent. The same skeptical students and Dr. Warren who booed Wells during the dentist's failed demonstration would be Morton's audience, also. Contained in Morton's contraption was ether, which had been developed several decades earlier.

On October 16, 1846, at the surgical theater in Massachusetts General Hospital, a nervous patient sat on a rusty chair. Medical students sat in the auditorium , awaiting his desperate cries when his neck tumor was carved out in moments. The restless Dr. Warren, with a scalpel in his hand, waited for the doctor who had promised to bring a pain killer. The man had not shown up.

Just as Dr. Warren was about to give up and prepare to operate on the patient's neck without the painkiller, Morton burst into the operating theater dragging a frazzled assistant behind him. The pair hurriedly moved towards the center of the operating room, carrying a peculiar-looking bottle with two tube-like portals. The flask contained a colorless liquid.

Unsure what to make of the scene, Dr. Warren nodded at Morton and said, "Doctor, your patient is ready."

After whispering a few reassuring words to the patient, Morton held one pipe of the bottle over a mask on the patient's mouth and asked him to take a few deep breaths. The patient did so, inhaling the vapors rising from the fluid in the flask. Soon the patient appeared to be relaxed and sleeping.

Morton looked back up at Warren and said, "Doctor, now *your* patient is ready."

With cautious hesitation but with swift moves, Warren removed the vascular tumor in about thirty minutes. The patient barely winced during the entire procedure. After the surgery Warren asked the patient if he had felt any pain. The patient said that he felt nothing.

Dr. Warren looked up at the audience and quietly whispered: "Gentlemen, *this* is no humbug."

Morton's demonstration was a glorious success.

This dramatic demonstration on an early fall day in Boston was the beginning of a convoluted, murky, and tragic story of claims and counter-claims of being the "first" person to discover surgical anesthesia.

Morton became an instant international celebrity, while Wells struggled in obscurity. No one believed the dentist's claim of being the first to demonstrate the power of nitrous oxide. Dejected and downcast, Wells quit his dental practice and became increasingly addicted to nitrous oxide and alcohol. Some years later he began abusing chloroform, which had just been discovered as an anesthetic agent.

In January of 1848, unable to bear the burden of his anguish, Horace Wells held his nose over a bottle of chloroform long enough to get its full effect. Just before collapsing, he forced a razor deep into his own thigh, severing the femoral artery and died.

About a week after his death, his widow Elizabeth Wells received an urgent letter from the Paris Medical Society (equivalent to the Nobel Committee of today). The academy—which had no knowledge of Wells' death—had reached its judgment: It was indeed Dr. Horace Wells who was the first to discover the anesthetic properties of laughing gas.

During the following century this reappraisal would be confirmed by other historians.

The surgical theater at Massachusetts General Hospital, where Morton demonstrated the effects of anesthesia, was named the "Ether Dome." On October 15, 1966, it was listed in the National Register of Historic Places in Boston.

A 40-foot granite structure was built at Boston's Public Gardens and dedicated on June 27, 1868. Through decades of negligence, lack of maintenance, and from the ravages of weather and vandals, the monument deteriorated. In 2000, a group of anesthesiologists initiated restoration efforts. With donations from the public and the City of Boston, the monument was restored to its original glory, and rededicated in 2006.

This monument to the conquest of surgical pain stands near the intersection of Arlington and Beacon streets in Boston's Public Gardens, not far from MGH. Named the *Ether Monument* (also *The Good Samaritan*), it is an imposing majestic structure.

Mankind should be grateful to Wells, Morton, Collins, and a few others for developing surgical anesthesia—one of man's greatest achievements.

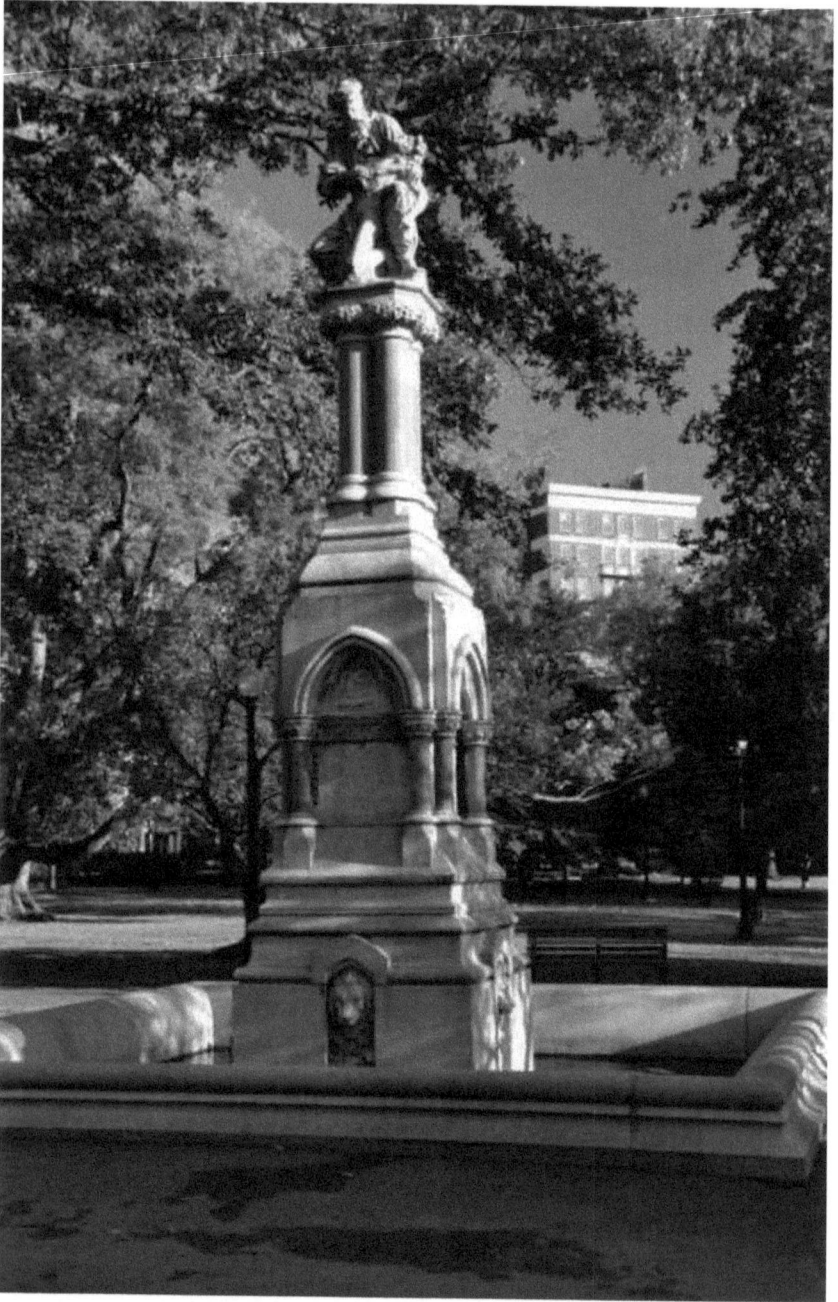

Ether Monument in Boston's Public Gardens

Ether Monument, close up.

A Mysterious Something

In the fall of 1921, four scientists discovered a "mysterious something" which continues to impact today's patients with diabetes. The scientists had extracted from dogs' pancreas a chemical that almost "cured" diabetes in animals, and in a handful of diabetic patients.

It was insulin—the first effective treatment for diabetes.

The phrase "diabetes mellitus" is derived from two sources: "diabetes," meaning "siphon" from Greek, and "mellitus," meaning "honey" from Latin. Together the word meant to say that the disease sucked up sugar from one's body, dramatically describing a major symptom of diabetes—passing large amounts of sugar in the urine. In India, diabetes was called "madhu meha," or "honey urine."

Other names for diabetes described the turmoil and suffering of its victims: "sweet sickness," "pissing evil," and "melting down of flesh and blood into urine."

Diabetes had been known since antiquity, but no treatment had been effective. Doctors resorted to bleeding, making blisters on the skin with red-hot iron rods, or prescribing huge amounts of opium so that dope "dulled the despair and eased the suffering."

Some doctors advised their patients to eat plenty of sugar to compensate for the loss of sugar in the urine, while others urged them to fast to prevent excess sugar accumulation in the blood, and to perform exercises to help burn the blood sugar. A French doctor remarked, "You shall earn your bread by the sweat of your brow," asserting that a diabetic patient should claim his meal only after an exercise that helped reduce his blood sugar.

Besides the lack of treatment, the cause of diabetes also remained mysterious until the mid 19th century.

Physiologists had discovered that the pancreas produced secretions that drained through tubes connected to the upper small intestine. From this it was surmised that the pancreatic juices were necessary for digestion of food. However, no one knew the chemical nature of these juices.

The German medical student Paul Langerhans (1847-1888) had taken up the study of the microscopic structure of the pancreas as part of his graduation work. He presented his thesis in 1869, in which he described that the pancreas had some peculiar features.

He noted that scattered throughout the pancreas, there were clusters of "clear cells," which when stained with dyes (to study the cells) took up different colors than the rest of pancreatic cells (called acinar cells). He also wrote that he did not know the functions of these clear cell clusters.

Langerhans' discovery of "clear cells" in the pancreas remained in obscurity for nearly 20 years, when the French scientist E. Laguesse rediscovered and named them the "islets of Langerhans." He suggested that these cells might be the source of an internal pancreatic secretion that was different from the usual secretions from the acinar cells that drained into the intestines to help in food digestion. By then it had been shown that a removal of the entire pancreas in the dogs led to lethal diabetes, but a partial removal of the pancreas did not cause diabetes.

Thus, at the beginning of the 20th century, scientists had accepted that the pancreas produced a *mysterious something*—a chemical— that probably came from the islets of Langerhans, which somehow controlled blood sugar. Neither the nature of the chemical nor the process by which it entered the blood and lowered blood sugar was known. The process of extracting the chemical, too, remained elusive.

Some scientists made crude extracts from the pancreas and fed them to diabetic patients, which turned out to be useless. We now know that in the crude pancreatic extracts, the digestive juices produced by the acinar cells of the pancreas destroyed the chemical manufactured by the islet cells of Langerhans.

Into this confused world of diabetes entered Frederick Banting (1891-1841), a surgeon with a fledgling private practice and a part-time lecturer's job at the University of Western Ontario in London, Canada.

One night, in preparation for his class the next day, Banting was reading a paper published in *Surgery, Gynecology, and Obstetrics*. The article described the autopsy findings of a man who had died from a complete obstruction of the pancreatic duct—the tube that carried the digestive juices from the pancreas into the intestine. The tube had been blocked from internal chemical stones, or the condition known as "pancreatic lithiasis."

The paper included histological pictures of the patient's pancreas. Looking at them, Banting was amazed to note that all of that patient's pancreatic acinar cells (which produced digestive juices) had been destroyed, but all the islets of Langerhans cell clusters were intact: they were the presumed source of the mysterious chemical controlling blood sugar.

A brilliant idea struck Banting. Why not experimentally occlude the pancreatic ducts in dogs so that the acinar cells are digested by their own secretions, while leaving intact the islets of Langerhans? Then one could extract the chemical from the leftover, undigested islet cells and test their effect in diabetic dogs.

But Banting was a practicing surgeon without laboratory facilities. His friends advised him to seek help from Professor John JR Macleod (1876-1935) at the University of Toronto, who not only had an established laboratory but also was an international authority in diabetes research.

Banting approached Macleod and proposed his research concept, but the professor was skeptical. He could not see how such a "simple idea" (even though Banting had called it "brilliant") coming from a practicing surgeon could work.

Banting was insistent; thus, in spite of misgivings, Macleod agreed to help him. Macleod allowed Banting access to his laboratory, let him use the dogs, and assigned Charles Best (1899-1978), a master's-degree student about to graduate, to assist Banting in his experiments during the summer months, when the laboratory would be relatively unoccupied.

From May through August of 1921, Banting and Best practically lived in the lab, conducting and refining experiments. Initially they were disappointed: many dogs died from infections. In some dogs, even after six weeks of pancreatic duct ligation, the pancreas remained healthy.

Banting and Best extract crude insulin from a dog's pancreas.

The first glimmer of success came on July 30, 1921. Upon receiving the crude islet cell extract, in the diabetic dog #401, the blood sugar concentrations dropped precipitously. However, the dog went into a coma and died the next day.

Banting and Best realized the historical importance of the results from this dog study: for the first time in history, they had shown that an extract from the islet cells *could* reduce blood sugar. Over the subsequent weeks, they performed more experiments, and were able to confirm their initial findings on several occasions. They then approached Professor Macleod with their results.

The professor, who had spent much of his summer vacationing at home in Scotland, looked at the results, but was not convinced. As an expert researcher, he could see many "holes" in the study design and procedures. However, he did not wish to disappoint the amateur scientists; he asked James Collip (1892-1965), an expert biochemist working in his lab, to try and see if a pure chemical could be extracted from the islet cells isolated by Banting and Best. Collip succeeded. He managed to produce enough of a relatively pure chemical, which could be injected into patients suffering from diabetes.

But, what should they call the chemical?

Banting and Best had proposed "isletin" to indicate the chemicals coming from the islet cells, but Collip proposed "insulin," since he felt it simple and easier to pronounce.

In February 1922, Leonard Thompson, a boy who was near death with severe diabetes at Toronto's Sick Children's Hospital, became the first human to receive insulin prepared by the four Canadian scientists. The first insulin preparation caused severe allergic reaction, but after receiving purified samples of insulin, Thompson regained his strength and appetite, making a spectacular recovery. The news of this discovery became an overnight sensation all over the world.

These scientists, among others, began intense collaboration with the pharmaceutical company Eli Lilley; within a year, large amounts of standardized insulin became available.

Banting and Macleod were awarded the 1923 Nobel Prize for their roles in the discovery of insulin—but Best had been left out of the honors, which made Banting very unhappy. He publicly announced that he was going to split his share of the Nobel Prize money with Best. Not to be outdone, Macleod followed suit, sharing *his* half of the money with Collip.

Controversies arose about the individual roles of these four scientists in the discovery of insulin, and missed recognition for others, especially for Nicolae Paulescu (1869-1961), a Romanian professor of physiology who had claimed that he was the first to show the blood sugar lowering properties of pancreatic extract in experimental animals.

However, Banting, Best, Macleod, and Collip's collective work led to the successful extraction and purification of insulin, enabled clinical testing, and to commercial manufacturing. Historians hail the discovery and development of insulin therapy as one of the greatest achievements in medical science.

The Miraculous Aspirin

Even though scientists in the 20th century might rightly boast about developing effective drugs and treatments unlike the past centuries, the highest honor for a great public health impact goes to an old, inexpensive drug—aspirin.

It is also fitting that the honor for developing the humble aspirin and for discovering the properties of its chemical predecessor, acetyl salicylic acid, will be shared by an ordinary chemist and a country reverend, respectively.

One day in 1758, the Rev. Edward (also known as Edmund) Stone (1702-1768) of Chipping Norton in Oxfordshire in England was on his daily stroll through the meadows. Of late he had been ill from fever or "agues"—a common term used in those days to describe all maladies that caused flu-like symptoms.

For no particular reason, he pulled and nibbled a small piece of twig from a willow tree. It was "extraordinarily bitter." To his surprise, however, his aches and fever were relieved soon thereafter. He knew that Peruvian Indians chewed the bark of the cinchona tree as treatment for malaria. Stone wondered if the bark of a willow tree might be good for all types of fevers.

Stone then prepared a dry powder from the willow tree bark and treated 50 patients who had fevers—the powder worked in all of them. On April 25, 1763, he wrote to the Royal Society about this discovery.

> *"As the willow tree delights in wet soil where agues [fevers] abound...so must it contain fever-curing properties...[because]... many natural maladies carry their cures along with them."*

As so often happens in history, this sensible conclusion was promptly dismissed by contemporary doctors.

Unknown to the reverend, willow tree bark, a folk remedy then, had a long history.

Two thousand years ago, Hippocrates, and the Egyptians 1,500 years before Hippocrates, had used many remedies from plant extracts. The decoctions of dried leaves of myrtle, poplar tree juices, and extracts of willow tree bark were used for fever, eye disease, and pain, especially labor pain.

We now know these extracts contain salicylic acid.

While the British disregarded the reverend's report, the Europeans did not. Amidst intense competition fueled by national pride, the French, Swiss, and Germans raced to find the fever-reducing ingredient in the reverend's bark.

In 1835, Karl Jacob Löwig had extracted an acid from the bark, which he called *spirsäure*. By 1838, the French had renamed it *l'acidesalicylique*, or salicylic acid. Others discovered that the oil of wintergreen and meadowsweet (*Spireaulmaria*) trees, too, yield rich quantities of salicylic acid.

Meanwhile, capitalizing on their dominance in the aniline dye industry, the Germans led the salicylic acid market, and it soon became the choice remedy for fever and pain all over the world. Matters stood that way until the 1890s.

A young chemist, Felix Hoffmann, working in the aniline dye department of the Bayer Division of Germany's I. G. Farben Company, was worried about his father's incapacitating arthritis. The senior Hoffmann had to consume six to eight grams of salicylic acid several times each day—a nauseating treatment causing excruciating abdominal pain.

Keen on finding a safer alternative to sodium salicylic acid, Felix experimented and soon hit upon a solution. He had prepared "acetylsalicylic acid," a minor chemical variant of the sodium salicylic acid. One night in 1895, he gave the first vial of the new compound to his father. It is said that the elder Hoffmann was delighted with this remedy because it provided his "first pain-free night" in years.

Aspirin was born.

In 1899, Hoffmann and Heinrich Dreser, a colleague at Bayer, coined the term "aspirin": "a" for acetyl and "spirin" from the German *Spiraure* for the meadowsweet tree of the *Spirea* family that yielded the salicylic acid. The drug was soon licensed, and the Germans once again claimed a huge share of aspirin's commercial success.

There were market-wars between the Bayer Company and other aspirin manufacturers about the patent issues. But, following the defeat of Germans in the First World War, the Bayer Company was forced to relinquish all its hold on both the name aspirin and its manufacturing patents.

Thus began aspirin's ascent in the U.S. It is estimated that in the U.S. alone, over 1 billion tons of aspirin are consumed annually.

The full glory of aspirin was yet to be realized. After the discovery of prostaglandins in the 1960s, John Vane (1927-2004) in 1971 hypothesized that aspirin blocked the prostaglandin action, demystifying the biochemical basis for aspirin's analgesic properties and anti-fever action. John Vane shared the 1982 Nobel Prize in Physiology or Medicine along with Sune K. Bergström (1916-2004), and Bengt I. Samuelsson (b 1934) for their discoveries concerning prostaglandins and related biochemical substances.

It is the anti-prostaglandin property of aspirin that is the basis for aspirin's modern use beyond headaches and fevers. From coronary artery disease and strokes, to preeclampsia, and clotting disorders, aspirin has become a mainstay of prophylaxis as well as the first-line therapy.

Thank you, Reverend Stone. Dankeschön, Herr Felix Hoffmann.

From Perkins Tractor to Goat's Testis

The first medical instrument patented in the United States in 1796 was a deceptively simple, but thoroughly useless device. It was the *Perkins Tractor* developed by one Dr. Elisha Perkins of Connecticut.

The instrument was no more than two metal rods, each measuring three inches long. One was made of a gold alloy and the other of silver. Perkins claimed that with his *tractor* he could "extract" and cure rheumatism, pleurisy, violent insanity, yellow fever, "inflammatory tumors," and many more ailments.

The sales pitch was so effective that President George Washington and his family used the Perkins Tractor, as did Chief Justice Oliver Ellsworth. When the public learned of this, the popularity of the device and its sales skyrocketed.

Obtaining patents to questionable medical devices did not just begin and end in the 18th century USA. Its continuation is as much a testimony to the cunning ingenuity of the device makers as it is to the gullibility of the poor public, eager for remedies for desperate conditions at any cost.

Other examples:

Patented tonics and drops made their appearance in the late 19th century. A $75 million industry, it flourished into the 1960s. Most of these were concoctions of medicinal alcohol, sweeteners, and hefty doses of cocaine and opium.

In 1885, the Lloyd Manufacturing Company of New York developed *Cocaine Toothache Drops,* which were sold at 15 cents a bottle. This was recommended especially for toddlers with teething problems.

Grove's Tasteless Chill Tonic was marketed for helping children gain weight. A 1920 banner advertising the tonic had a picture of a pig with the head of a smiling baby. It proclaimed that the tonic *Makes Children and Adults as Fat as Pigs.* At 50 cents a bottle—expensive those days—over 1½ million bottles were sold.

The *Switch Engine* mixture purporting to cure "all sorts" of intestinal troubles sold at 25 cents a bottle. "The solution is especially useful," claimed its manufacturer, "to keeping the Main-Line Clear." But there were skeptics. One disgruntled customer complained dryly that the remedy "opened men's purses by opening their bowels."

Page's Inhaler, a product for people with asthma, was really dangerous. It contained 100 "special" cigarettes for asthmatics so that they could "enjoy the palliative effects of smoking." It sold at $1.20 a box.

As technology advanced in the 20th century, so did quackery, with incorporation of machine-age concepts into their products. Dubious manufacturers used electricity, x-rays, radio waves, electromagnetism, solar power, and computers to perfection, proclaiming the benefits of hundreds of "miraculous machines."

In the 1930s, one Mr. Dhinshah Ghadiali developed the *Spectro-Chrome.* It was a room filled with colored lights, in which his patients were made to sit naked. The lights were supposed to cure illnesses ranging from migraine to malaise, from arthritis to anorexia. The *Spectro-Chrome* became a cult classic, garnering $1 million in profits for the inventor. The show ended, however, in 1947, when the FDA prosecuted Ghadiali on charges of fraud. He was found guilty on 12 criminal counts.

Albert Abrams invented the *Dynamizer,* a ludicrous assembly of electrodes and tubes, used to detect diseases using a single drop of blood collected from the patient as he faced west. Abrams injected the blood into a tube in the machine, and connected the machine to a healthy volunteer. By pressing the abdomen of the healthy volunteer, Abrams claimed that he was able to "read" the exact diagnosis of the patient whose blood had been taken. He also designed a smaller version, the *Microdynameter*, as a tool to "measure" electricity in the body and to diagnose almost all diseases.

In 1962, the courts banned the selling of all of Abrams' devices. The developers of the *Recto-Rotor* also made fantastic claims. The device consisted of a six-inch tube, shaped like a blunt candlestick, connected to an electrical motor. The subject inserted the tube through his anus and turned on the motor to stimulate the rectum to "cure all ailments." The inventors claimed that the instrument worked because it was the "only device that reached the Vital Spot."

Quacks continue to strike the most vulnerable segments of our society. The elderly with arthritis, cancer, insomnia are especially susceptible, as are those suffering from obesity who want to lose weight. No wonder, then, that the weight-loss industry is replete with remedies promising melting away of excess fat with undeniable certainty.

In 1995, the U.S. Food and Drug Administration (FDA) restricted the sale of one such device. The *Acu-Stop 2000* was an instrument designed to be inserted into one's ears like a hearing aid. The manufacturers claimed that when activated, the device would stimulate one's "acupressure points," suppressing appetite.

Nothing can match the ingenuity of Dr. John Brinkley (1885-1942) of Kansas. Observing that goats did not lose sexual drive as they got older, but men did, he began grafting goats' testicles into the sacs of "non-functioning glands" of his patients. By the time his medical license was revoked for malpractice, Brinkley claimed that he had enhanced the carnal performances of more than 16,000 men. No independent verification of his statistics exists; but one satisfied customer boasted to have been on Brinkley's success list. He proclaimed that by the grace of the goat's glands, he had fathered a son within one year of the transplant. He had gratefully named his poor son Billy.

Why do perfectly normal, sane, and competent people fall into traps and buy such products?

As serious and debilitating illnesses take their tolls, hope remains eternal. Even the hardiest amongst us hope for cures, however meaningless the remedy appears. Bob McCoy, the curator of the Minneapolis' Museum of Questionable Medical Devices, one of the largest museums of its kind, reasons, "...because most of us want to believe in miracles."

A Gift from the Mother Earth

In 1882, Robert Koch announced that a tiny, rod-shaped bacterium was the "Captain of Men of all Death" —the agent that caused tuberculosis. Yet, an effective treatment for tuberculosis did not materialize for seven more decades.

When it came, it was a gift from the mother earth.

Selman Abraham Waksman (1888-1973), born in Priluka near Kiev, immigrated to the United States in 1910 and studied soil bacteriology at the Agriculture College of Rutgers University. He wanted to understand the process by which waste products were degraded in the soil. Waksman studied soil samples from the campus at Rutgers and, after many years of research, discovered that soils contained numerous microbial florae, which were responsible for degrading animal and plant waste products producing humus and peat.

By 1939, René Dubos (1901-1982), Waksman's postdoctoral student, had extracted from soil bacterium *Bacillus brevis,* two germicides that were effective against bacterial infections in cattle. The germicides—named "tyrocidine" and "gramicidin"—were, however, too toxic for humans.

Waksman pursued the line of research Dubos had initiated, searching for chemicals from the microbes in the soil that may be effective against human infections. In 1944, Albert Schatz (1922-

2005) and Elizabeth Bugie (1921-2011), two other graduate students of his, isolated *Actinomyces (streptomyces) griseus* from the throat of infected chickens. From this fungus, the team extracted a new chemical agent that killed some strains of gram-negative bacteria.

Waksman then showed that this chemical also killed *Mycobacterium tuberculosis*, the microbe that caused tuberculosis. He named the compound "streptomycin," and christened the group of such chemical agents, "antibiotics."

Waksman supplied ten grams of streptomycin to William Feldman and H. Corwin Hinshaw at the Mayo Clinic, who confirmed its anti-tuberculous properties. The sensational news of their findings was received warmly by the medical world, for it was the first time a drug had been shown to kill the TB bacterium. Soon, large clinical trials were designed in Great Britain establishing the efficacy of streptomycin against tuberculosis.

Waksman persuaded the Merck Company to manufacture streptomycin. Merck reaped huge profits from streptomycin sales, with 80% of royalties going to Rutgers, and 20% to Waksman. In 1949, following a litigation that was settled out of court, Waksman shared the financial gains with 28 of his students, including Schatz.

Waksman was the first director of Rutgers' new Institute of Microbiology, which was built from the streptomycin profits.

Waksman was awarded the 1952 Nobel Prize in Physiology or Medicine for the discovery of streptomycin. He remained professionally active all his life, writing over 20 books, including *My Life with the Microbes* in 1954 and *The Conquest of Tuberculosis* in 1965.

Despite a great first step in developing streptomycin and other anti-tuberculosis drugs, the battle against TB is far from complete; unlike smallpox, TB remains to be conquered. In parts of Africa and Asia, scientists are battling this old menace that now has a new face—strains of tuberculosis bacteria that are multi-drug resistant.

Selman Waksman

Some Like it Sour

As the magnificent sailing vessels in the 17th century floated across the oceans in search of conquest and commerce, the soldiers, sailors, and slaves working in those vessels were being tormented with unspeakable maladies. Scurvy, the "plague of the sea," was their foremost enemy.

The symptoms of scurvy would begin slowly with bleeding in the skin and gums. The progression of symptoms was, however, rapid: nervousness, "degeneration of spirits," and exhaustion would lead to agonizing death within days of onset of the first symptoms. In a 1740-44 voyage around the world led by Lord Anson, 761 sailors died of scurvy—only 200 returned home alive. During the age of naval expeditions of the 17th through the 19th century, more than one million seamen had died of scurvy.

In 1747, young James Lind (1716-1794) a surgeon's apprentice, carried out the world's first "controlled clinical trial" and changed British naval history.

Lind was born in Edinburgh and joined the military service early. His duties were to nurse ailing soldiers, document their illnesses, and report them to higher authorities.

During voyages to Africa and the West Indies, Lind saw the fury of scurvy firsthand. He was also appalled at the horrible living conditions for sailors and the disgusting nature of food they had to eat. They slept in decks infested with rats and vermin, and ate rations of mold-covered pork, maggot-laden meat, and withered chunks of dry biscuits. It was known that fresh vegetables and fruits were healthy—but no one had seriously considered translating this knowledge into practice, for it was quackery that dictated medical care and eating choices.

Lind carried out a simple experiment on sailors with symptoms of scurvy. In his book *Treatise on the Scurvy* published in 1753, Lind wrote:

> *"On the 20th of May... I took twelve patients [of] scurvy on board the Salisbury at sea. They all...had putrid gums, lassitude, with weakness of their knees. Two were ordered [to drink] each, a quart of cider a day; two others, twenty-five drops of elixir vitrial, three-times-a-day; two others, two spoonful of vinegar three-times-a-day; two of the worst patients were given one-half pint-a-day of sea water; two others, each [ate] two oranges and one lemon every day, and the remaining two [were given a concoction of] nutmeg... garlic, mustard seed, balsam of Peru, and gum myrrh.*

> *"The consequence was that the most sudden and visible good effects were perceived from the use of orange and lemon...one [sailor] was fit for duty by six days."*

Lind also found that cider had an intermediate effect and that all the other treatments were useless. By 1753, when his book was published, Lind had retired from the navy.

He wrote other books, considered classics: *Essays on Jail Distemper,* and *On the Most Effectual Means of Preserving the Health of Seamen.* In these, Lind recommended steps to improve hygiene and prevent typhus among seamen.

But, Lind's remedy for scurvy, though very simple, was not adopted by the military until 1795—one year after his death. At the urgings of Sir Gilbert Blane, a follower of Lind, the British Navy finally mandated lemon juice as part of sailors' diet and, in one stroke, wiped out scurvy among its sailors.

But this ruling also earned the British navy and its soldiers the derogatory nickname *limeys*.

Casimir Funk (1884-1967), a Polish scientist, coined the term "vitamine" in 1912 to describe the chemicals as the "vital amines" found in foodstuffs. Since they were not all amines, the term was later changed to "vitamin."

In the 1920s, Albert Szent-Gyorgy (1893-1986) isolated "anti-scorbic acid" (later called "ascorbic acid"), which was the anti-scurvy chemical in the orange, cabbage, and of course, Lind's lemon juice. Szent-Gyorgy was awarded the 1937 Nobel Prize in Medicine for this discovery. That year's Chemistry Prize went to Walter Norman Haworth (1883-1950) and Paul Karrer (1889-1971) for synthesizing vitamin C.

The era of modern nutritional science Lind had initiated nearly 200 years earlier had just begun.

In 1753, James Lind demonstrate eating orange and lemon will cure scurvy. Many years later, the British Navy adapts using lemon among seamen to prevent and treat scurvy.

Vital Link

What disease causes burning palms, swollen feet, pounding hearts, breaking bones, putrid gums, and scaly skin?

These are symptoms of deficiencies of tiny amounts of essential nutrients our bodies cannot make—the vitamins.

The complexes of symptoms of micronutrient deficiency have been known since the time of the Pharaohs, but their causes remained a mystery until about 100 years ago. Thanks to accidental, if surprising, discoveries by a Dutch army officer, a British biochemist, and a Polish American, most of the serious cases of vitamin deficiencies have all but disappeared.

The best known of these was beriberi, a killer that affected large populations in the Far East. Symptoms of beriberi included fatigue, swelling of the hands and feet, confusion, and personality changes. These led to death from intractable heart failure. The disease seemed to manifest with frightful regularity and came in waves of epidemics, because of which beriberi was considered to be a contagious diseases caused by microorganisms.

In 1886, the Dutch government sent a commission to its colony in Java to discover the cause of beriberi and save its soldiers. The group invited a young doctor, Christian Eijkman (1858-1930), on leave from the Dutch East Indies army to join them. The group's efforts did not succeed in finding the cause for beriberi, and the commission returned home, but Eijkman stayed in Jakarta pursuing his quest for the "beriberi-causing germ."

He chose to conduct his studies in chickens because they were cheap and it was known that beriberi killed chickens in large numbers. By injecting the "infected" materials from animals with beriberi to healthy chickens, he hoped to transmit the disease.

In 1889, a major epidemic of beriberi killed nearly three-fourths of all animals, including the non-injected (control) chickens. However, as mysteriously as it had started, the epidemic suddenly vanished.

Upon investigating further, Eijkman learned that about three weeks before the beriberi epidemic, a servant had used table scraps of rice from the army officers' mess to feed the chickens. Sometime later, a new cook refused to let the military rice be fed to the civilian chickens and had switched to cheaper, unpolished rice.

Eijkman recognized that the "military rice" caused hens' beriberi and the cheaper, unpolished rice had cured it. Initially Eijkman thought that the unpolished rice had a factor that countered some "toxins" in the polished rice. He then experimented on prisoners. He fed them polished rice and then switched back to unpolished rice. The former diet caused beriberi, while the latter cured it. Eijkman concluded that raw rice had an "anti-beriberi factor," which was destroyed by polishing.

Frederick Hopkins (1861-1947) of London, working at Cambridge's Trinity College, was the head of the world's first biochemistry department. A brilliant scientist, he had discovered the amino acids tryptophan and glutathione, and was studying their roles in the diet. When he fed artificial combinations of food to laboratory animals, Hopkins found that although the diets contained adequate calories and had all of the essential amino acids, proteins, carbohydrates, fats, and salt, the animals became severely growth-retarded. By the additions of small amounts of milk, however, he could restore their growth. Based on these findings, in 1912, Hopkins concluded that natural foods contained "essential, growth-promoting, accessory factors" absent in synthetic, artificial foods.

At about the same time in London, Casimir Funk (1884-1967),

a biochemist born in Poland, was studying the chemistry of rice polishing. He isolated a chemical which he named "thiamin," with which he cured beriberi in laboratory pigeons.

In 1912, Funk coined the term "vitamine" because he thought thiamin and similar food chemicals were chemically "amines," vital for survival. Funk also isolated niacin, but did not realize its nutritional value at the time. Three years later he sailed to the US and spent the rest of his illustrious career in New York.

By the early 1920s, the stage was set to solve the mystery of the ancient nutritional diseases.

Eijkman's anti-beriberi factor was shown to be thiamin, which Funk had discovered: "vitamines" were shown to be Hopkins' "essential accessory factors." Later studies showed that not all of these compounds were "amines," and hence the letter "e" was dropped—the generic term "vitamin" was used for all organic micronutrients the body was unable to make. By the mid 20th century, over 20 vitamins had been discovered, and a multibillion-dollar industry had begun to boom.

While three biochemists worked on the same topic, their fortunes varied. Funk received numerous honors, including an institute of research named after him. In 1992, the Polish government honored him posthumously by issuing a postal stamp with his likeness. But Funk did not receive the big prize—Eijkman and Hopkins shared the 1929 Nobel Prize for the discovery of the anti-beriberi factor and vitamins.

The Saint and the Sleeping Pill

The chemical *barbituric acid* is the core compound for barbiturate preparations and was named after a saint, St. Barbara, for no apparent scientific reason.

According to legends, Barbara was the daughter of Dioscorus, a 3rd century (CE) man who lived in a region now in Syria. Dioscorus was jealous of his daughter's beauty, and so he confined her to a tower to shield her from outsiders. Barbara was attracted to Christ's teachings, and once when her father was away, she converted to Christianity. When he learned this, Dioscorus became furious. He had Barbara dragged into the city and tried by the prefect of the province. She was condemned to be beheaded, a punishment Dioscorus himself carried out. Not long thereafter, lightning struck him, reducing him to ashes.

The legend of lightning in this story was the reason for the faith linking St. Barbara as a savior from fire, thunderstorms, and natural disasters. She became the patron saint for firefighters, artillerymen, and miners, who face such dangers regularly.

But, why is she connected to barbiturates?

On December 4th, 1864, a German chemist, Adolph von Baeyer (1835-1917), who later founded the Baeyer drug company, discovered a compound "melonylurea" derived from a reaction of urea and malonic acid. To celebrate the discovery of this new compound he went to a local tavern, where artillerymen and miners were also celebrating *St. Barbara's Feast Day* on December 4th.

In those days, professional groups celebrated their respective patron saint's Feast Day with songs, dance, and award ceremonies. As he enjoyed the St. Barbara's Feast Day, Baeyer thought of naming the new compound he had just discovered, *barbituric acid"*—"barb" for Saint Barbara, and "uric acid" because of the compound's origin from urea. Neither he nor his colleagues knew of the potential therapeutic value of barbituric acid at that time.

That discovery came 40 years later.

In 1903, Hermann Emil Fischer (1852-1919) and Joseph von Mering (1949-1908) prepared a compound from barbituric acid by substituting two ethyl groups for two hydrogen molecules attached to carbon. They soon discovered that the new chemical induced sleep in human subjects.

At that time, von Mering lived in Verona, Italy—hence he named the chemical *Veronal*. Verona also means "that which causes a sensation of peace and solace." Von Mering believed that Verona was the most peaceful city in the world, a good fit for the drug that seemed to induce a peaceful sleep.

Veronal was the first barbiturate to be sold as a sedative and, sadly, also became one of the most widely abused, habit-forming drugs.

In 1912, Luminal (phenobarbitol), a major variant of Veronal, was developed. Besides producing deep sleep, Luminal was shown to contain potent anticonvulsant (anti-seizure, or anti-epileptic) activity.

Other anti-epileptic drugs that were developed over the next two decades were: Amytal (amobarbital); Nembutal (pentobarbital); Seconal (secobarbital); and many short-acting barbiturates, such as Evipal (hexobarbital); Pentothal (thiopental), and Brevital (methohexital). All these were derivatives of barbituric acid. Therefore, this class of drugs came to be called the *barbiturates*.

The discovery of barbiturates and the research into their pharmacology and chemistry also led to the discovery of the highly useful benzodiazepines, such as Valium (diazepam) and Halcion (triazolam).

In 1902, Fischer won the Nobel Prize in Chemistry. The citation stated that this was *"in recognition of the extraordinary services he has rendered by his work on sugar and purine syntheses."* Three years later, Fischer's organic chemistry teacher followed suit.

Adolph von Baeyer won the 1905 Nobel Prize in Chemistry, "in recognition of his services in the advancement of organic chemistry and the chemical industry, through his work on organic dyes and hydroaromatic compounds."

Fischer and Baeyer did not consider their roles in the development of barbiturates as being very important—they felt that this was the inevitable byproduct of their enduring interest in organic chemistry.

Two Controlled Trials on Bloodletting

In 1946, Sir Bradford Hill in England designed and conducted a randomized controlled trial (RCT) funded by the UK's Medical Research Council (MRC). Often this has been called the first randomized control trial conducted in modern times.

Sir Bradford tested the effectiveness of the antibiotic streptomycin in patients with pulmonary tuberculosis. The study was unique in many respects, including "masking" of treatment assignment (or double blinding) to avoid investigator and patient bias. The study proved that streptomycin was indeed effective in curing TB—the first modern drug to show that TB can be cured.

As major as the MRC streptomycin trial was, it was not really the first RCT. There are at least two older, if bizarre, examples from medical history of trials that incorporated the idea for balancing treatments and placebos.

Van Helmont's Challenge

John Baptista Van Helmont (1579-1644), a Flemish physician, questioned the value of bloodletting and excessive purging—common practices to cure all sorts of illnesses. He challenged his contemporary colleagues known as "humoralists," who believed in bloodletting. Von Helmont's challenge, paraphrased in today's language, would read:

"If you say that you can cure any kind of fever without evacuation [purging] come down for a contest. ...

"Let us take 200 to 500 poor people with fever, pleurisy, etc., from a hospital, from a camp, or from elsewhere. Let us divide them in halves. By casting lots, one half of them will be my share and the other half yours. I will cure them without bloodletting and [with] only sensible evacuation; but you can treat them as you wish, including bloodletting.

"There shall be a wager of 300 florins deposited by both parties [for the winner to take all]. We shall see how many funerals both of us shall have."

It is not clear if anyone took up this challenge and risked losing 300 florins; but Van Helmont had a vision of comparing likes with likes, and randomly assigning the treatment groups.

Alexander Hamilton's Thesis

Alexander Hamilton, an early 19th century doctor, was required to write a thesis to earn his MD degree from the Edinburgh University. In 1814, he presented the thesis and provided the details of a clinical trial he had designed and carried out with the help of two army surgeons. It seems the "trial" was carried out while they were engaged in the Peninsular War against the French.

"It had been so arranged that this number [366 sick soldiers] was admitted alternately in such a manner that each of us had one third of the whole. The sick were indiscriminately received, and were attended as nearly as possible with the same care and accommodated with the same comforts. One third of the whole

*were soldiers of the 61st Regiment, and the remainder of my own,
the 42nd Regiment. Neither Mr. Anderson [the second surgeon]
nor I ever once employed the lancet. He lost two, and I four cases;
whilst out of the one third [treated with bloodletting by the third
surgeon] thirty-five patients died."*

An impressive result from *not* doing what was considered a standard treatment for all sorts of illnesses. Only 5/244 had died among the group in which bloodletting was not carried out, for a mortality rate of 2%. By contrast, 35/122 had died among those who received bloodletting, for a mortality rate of 29%. No need for modern statistical analysis to prove that the results were significant.

However, there are no independent verifications that this study was ever carried out. But, even if this was an imaginary trial conceived to impress his MD examiners, Hamilton's concepts were evidently brilliant. He had grasped the essence of avoiding recruitment bias, and assuring the comparability of other treatments.

Hamilton did not publish his findings; but the results he presented in the thesis suggest that he had no faith in bloodletting or purging as beneficial treatments for any ailment.

But his contemporary doctors probably did not read his thesis or believe in his conclusions. Bloodletting and indiscriminate purging remained in vogue for another 75 years.

The Man Who Loved the Sun

In Denmark's Faroe Islands, the summers are cool and the winters are mild. But sunny days are rare. Strong winds and rainstorms blast the islands throughout the year.

Growing up in Tórshavn, Faroe Islands, young boy Niels Finsen (1860-1904) longed for the sun and savored its energy whenever he could. He suffered from a rare form of connective tissue disease (diagnosed as Pick's disease), because of which he always felt exhausted. Under the sun his body felt stronger and spirits uplifted. As he grew up, Finsen was struck by the effects of sunshine on the minds and bodies of humans and animals.

He was also puzzled: What is the sunlight made of? Can it heal? Can it hurt? How can we harvest the sun's energy?

After getting a medical degree at Copenhagen University, Finsen began conducting research on sunlight. He studied its effect on laboratory animals, and in the 1890s he discovered that sunlight killed bacteria and other microbes over the skin.

He tried sunlight on patients with smallpox skin lesions, which did seem to help them. After a long series of experiments, Finsen developed "light therapy" for "lupus vulgaris," literally, "a vulgar wolf" —a horrible skin condition caused by tuberculosis with serious disfigurement and pain.

With help from other scientists, Finsen invented the "Finsen-

Ryan Lamp," a torch containing a carbon arc light source for focusing on the skin with many lenses. He built a Phototherapy Institute, or a "Light House" for patients with skin tuberculosis. Over 800 lupus vulgaris patients were treated in the first year alone with 90% cure rate.

Finsen became world famous. In 1903, he won the Nobel Prize in Physiology of Medicine for developing the first effective treatment for skin tuberculosis. However, he was too ill to go to Stockholm to receive the award, and the next year he died at 44.

Finsen's fame continued to increase even after his untimely death. Now there is a medical school named after him, and a Finsen stamp issued by the Danish government. Over the years, phototherapy became a standard tool for treatment of skin ailments.

Due to Finsen's discovery, a "School of the Sun" was built on the European Alps. Thomas Mann, a famous TB victim, was treated here. Mann used the mountain backdrop in his novel *The Magic Mountain*.

However, the magic cure for TB—an antibiotic—did not materialize until the 1950s.

What was the illness Finsen had all his life? It was probably not Pick's disease, as we know it today. This condition has signs and symptoms of brain degeneration, which Finsen clearly did not have. In a 1935 Canadian Medical Journal paper, a scientist presented the findings of "Pick's disease" and called it *mediastino-pericarditic pseudo-cirrhosis*—a long title meaning a condition with inflammation in the chest and over the outer layer of the heart, constricting the heart, which can lead to liver failure. All of these can occur from common strep-throat infection leading to rheumatic fever and rheumatic heart disease—thankfully, very rare these days.

Prostate

Some historians think that the word prostate comes from the Greek root "pro" meaning "in front of," implying that it "prostates," or "stands as a leader in front" of the bladder.

But the prostate surrounds the upper urethra, the tube that carries urine out of the bladder, and hence sits *below* the bladder not in front of it. Therefore, the name given to it by the Greek anatomist Herophilus seems more appropriate: he called it *"prostate adenoidae,"* which means the "organ standing in front of the gland," the testicles.

Although the prostate was known for a long time, its functions were not understood until recently; often even the diseases of prostate were mixed up with those of the bladder. These included those leading to obstruction of the urine flow.

During much of ancient history, the inability to pass urine was an important, often the only reason to see a surgeon. The most frequent cause of this symptom was obstruction of the tube exiting the bladder by bladder stones—chemical structures resulting from imbalance of urine composition, often superimposed by infections.

Surgeons who specialized in removing bladder stones called themselves "lithotomists," or "stone smashers." In the 17th century, "wandering lithotomists," or surgeons looking for people with bladder problems, traveled European cities, promising patients an instant surgical cure. Since there were no anesthetic drugs, antiseptic agents, or antibiotics, most lithotomies were brutal acts of kindness only the most desperate would undergo.

A typical lithotomy would go like this: The patient would be tied down with his legs held up and wide by two strong assistants. The lithotomist then made a deep cut in the patient's perineum, inserted his fingers, and scooped out whatever he touched. The surgeon hoped to feel a tumor or a stone at the mouth of the bladder, and could scoop it out. When this did occur, the patient's symptoms of urine obstruction would be relieved. Sometimes the surgery worked—most often it did not.

The procedure appears barbaric now, but this *was* the standard of care. It is great that we have come a long way.

Baby Shots

He was not a king. He was never elected to an office. But when he died, the whole country wept. Its citizens lined up by the thousands along his funeral route, and the president of the Republic attended his memorial service. The newspapers worldwide eulogized him.

Louis Pasteur (1822-95), the man who would be the king of sciences, was the son of a humble tanner in Dôle, a village near the French Alps. He studied science and earned a doctorate from the prestigious École Normale in Paris. In 1834, he became a professor and was later appointed dean of the newly organized Faculty of Science at Lille, in the heart of French wine country.

The owners of the winery faced financial disaster. Their great wines were turning sour for unexplained reasons. To save their industry, they sought Pasteur's help to solve the enigma of sour grapes.

After a series of brilliant experiments, Pasteur discovered that when yeast was added to the liquid preparations of beet and sugar, it fermented naturally into a perfect brew, while bacterial contamination produced lactic and acetic acids, spoiling the wine. Pasteur suggested a simple remedy: heat the fermenting solution to $55°$ C prior to brewing. Thus was born the process of "pasteurization," which saved both the French wineries and the world's food-processing industry.

Pasteur then asked a profound question: how did the bacteria that putrefied wines enter the containers of fermenting brew?

Scientists had been promoting the notion that maggots in decaying organic materials appeared spontaneously—and they called the phenomenon "spontaneous generation." But Pasteur doubted this hypothesis.

To study this, he designed a glass flask with an "s"-shaped neck that could filter off air by trapping germs at the neck of the flask. He found that when organic solvents were stored in the new flasks, the chemicals remained pure, whereas if stored in regular flasks the chemicals putrefied. This was the first hint proving that there was no such thing as "spontaneous generation"—all putrefaction must arise from contamination from germs in the air.

As he was understanding his new ideas about germ pollution, Pasteur was approached by the silk industry leaders, who asked him to solve the mystery of "pébrine," a lethal disease of silkworms causing huge losses for French silk production.

Pasteur's studies led him to conclude that bacterial contamination of moths and their eggs led to the disease in the silkworms; by selective breeding with disease-free eggs, one could prevent the epidemic. Once again, he had saved the luster of the French silk industry.

Pasteur then generalized the concept that diseases can be caused by microorganisms invisible to the naked eye, giving rise to "the germ theory of disease." He proceeded to show how germs were responsible for anthrax and chicken cholera, among other infections. By the late 1870s, Pasteur was a household name in France—but it was his research in rabies that brought him universal admiration.

Using tissues from rabies-infected animals, Pasteur first discovered that rabies attacked the nervous tissue. He made cultures from these tissues and extracted an attenuated form of rabies virus; he hoped to test its efficacy, although he was not sure if it would work.

On July 6, 1885, Joseph Meister, a victim of a vicious dog bite, was brought to Pasteur's laboratory; the boy's mother pleaded with the great scientist to try anything to save her son from what would be an inevitable death. Reluctantly, Pasteur inoculated the boy with the new rabies vaccine; Joseph survived.

Louis Pasteur supervises administering anti-rabis vaccine.

The news was an international sensation. Despite initial failures, the vaccine would save thousands of rabies victims.

In 1888, an institute was built in Paris dedicated to the research and treatment of infectious diseases and was named the Pasteur Institute, which remains one of the preeminent research institutes in the world. It was here in the 1980s, the AIDS virus was isolated.

When the Nazis occupied Paris in 1940, they were about to desecrate Pasteur's burial crypt. Unable to witness this tragedy, Joseph Meister, who had become a gatekeeper at the Pasteur Institute, took his own life.

Clean Hands Ignác

IgnácSemmelweis (1818-1865) was misunderstood and ignored in life, and canonized and declared "defender of motherhood" after death.

Ignác (also spelt *Ignaz*) was born into a merchant family in Tabán, a section of today's Budapest. He graduated in medicine from the University of Vienna in 1844, and while waiting for a permanent job at the Lying-in Hospital in Vienna, he worked under many professors; as part of his duties, he needed to conduct autopsies.

While performing autopsies on women who had died during childbirth, he could see ravages of "childbed fever" (puerperal sepsis) in all its gory details. No one knew why puerperal sepsis occurred.

Semmelweis worked as an assistant to Professor Dr. Johann Klein (1788-1856) in 1846. He noted the very high death rates among women in Dr. Klein's First Obstetric ward, where only doctors and medical students performed deliveries. In the Second Division, where only midwifes performed deliveries, mortality rates were much lower.

This was common knowledge: poor Viennese women would tearfully beg to be admitted to the Second Division rather than to the First.

Young Semmelweis made an extensive study of the mortality patterns and became convinced that the riddle of childbed fever should be solved through understanding the cause of mortality differentials.

He noted that there was one striking difference in practice:

The doctors and students in the First Division began their day's work by working in the morgue performing autopsies on women who had died a few days earlier. Midwifes in the Second Division were not required to perform autopsies. Semmelweis also compared the autopsy findings of newborns who had died along with their mothers, and found the pathology to be identical. He thus concluded that childbed fever was not just a women's disease—even children died from it.

A tragedy in March 1847 jolted Semmelweis. Dr. Jakob Kolletschka, a much admired professor of forensic pathology, died of infection following an accidental finger wound while performing an autopsy. Semmelweis mourned his professor's death, and read the autopsy findings of the professor. He was astonished to find similarities in organ pathology found in his professor's body to those of women dying from childbed fever.

Later Semmelweis would write:

"Totally shattered, I brooded over the case [of his professor]... suddenly a thought crossed my mind: childbed fever and the death of Professor Kolletschka were one and the same...His...sepsis and childbed fever must originate from the same source. [The cause of this] was to be found in the fingers and hands of students and doctors, soiled by recent dissections... [and they] ...carry those death-dealing cadavers' poisons into the genital organs of women in childbirth..."

Now he focused on getting rid of the stench and "death-dealing" poisonous materials from his hands by washing them with many cleansing agents. Eventually he settled on a solution of chlorinated lime.

IgnácSemmelweis

In May 1847, notices were posted in the First Division "...All students or doctors who enter the wards must wash their hands...in a solution of chlorinated lime..."

The results were astounding.

The death rate declined dramatically. By May of 1848, the first full year of implementing hand-washing policy, maternal death rate had dropped to an unprecedented 1.27%.

As Semmelweis celebrated this success, his colleagues ridiculed him. Professor Klein was furious about the expensive and "outrageous extravagances" of using chlorinated lime water and fresh sheets for *every* patient. Medical students grumbled at the frequency of hand-washing; those currying favor with Dr. Klein even sabotaged the disinfection efforts.

From this time onward, Semmelweis seemed to be haunted by misfortune. His contract was not renewed in 1849, forcing him to move to his home in Pest, where he joined the University of Pest faculty.

But in Pest, Semmelweis' concepts about childbed fever were not received warmly. This was in part because Semmelweis was tardy in publishing the results of his research. He presented his first formal paper on puerperal sepsis in 1850—full three years after the discovery—at the Viennese Medical Society. It took him 11 more years to publish the report entitled, *The Cause, Concept, and Prophylaxis of Puerperal Fever.* Even after this 1861 paper, his contemporaries were not willing to substantially alter their opinion about Semmelweis' ideas concerning puerperal sepsis and his proposed means to prevent it.

In addition to the delay in publishing, there were other reasons for the poor reception of Semmelweis' theories by his contemporaries. His long and rambling writing appeared illogical and difficult to understand; the tone of his writing was one of extreme combativeness; and the microbial origin of infections was unknown during his time. Thus, even for those who were sympathetic to Semmelweis, it seemed inconceivable how a simple act of hand-washing could prevent such a horrible disease as childbed fever.

Angered by the poor reception of his discoveries, Semmelweis began writing scathing, "open letters" to professors who doubted his theory. He identified them by name and cried out that they were "murderers" and "medical Neros" full of ignorance.

In his final years, life became even more unbearable for Semmelweis. According to recent research, he probably developed Alzheimer's pre-senile dementia, a condition in which in addition to amnesia, the patient suffers from severe bouts of depression alternating with bouts of extreme euphoria. In July 1865, his family committed him to an asylum in Vienna, where he died two weeks later—he was 57.

Semmelweis died young and could not see his theory vindicated by such luminaries as Louis Pasteur and Joseph Lister (1827-1912).

What is the status of hand-washing to prevent infections today?

More than 150 years since Semmelweis' demonstration of its value, health care professionals continue to ignore the simple practice of good hand-washing before and after touching and handling their patients even today. In some studies as few as 14% of doctors and 25% of nurses washed their hands prior to handling patients.

Sir Joseph Lister confirmed Semmelweis' concepts.

Two Deliveries that Changed World History

In 1537, a 30-year-old woman went into labor to deliver her first child. Even after 24 hours of labor, her cervix had dilated only five centimeters, and the head of the fetus was too high in the pelvis. The mother's uterus was contracting poorly.

The doctors had two options: cut open the fetal skull to deliver the baby and save the mother, or save the baby by doing the life-threatening cesarean operation, seldom done on living women in those days.

The father made his choice clear. He is quoted to have said, "By any means, save the child, for another wife can be found easily."

When the father is the King of England, what choices did the doctors have?

Two days later, a puny baby was born in secrecy. How exactly Jane Seymour, wife of King Henry VIII, was delivered of her baby is shrouded in mystery. But what is known is that the much beloved queen died 12 days later, ostensibly from infections inside her belly.

Poets and commoners eulogized the queen and secretly blamed the king for forcing a cesarean section and a medical "murder."

The imperial offspring, who almost died as a fetus, however, became King Edward VI. During his brief reign, Protestantism became firmly rooted in England, forever changing the course of the history of England.

The second story concerning the birth of the future King of England is more tragic.

Four of the best doctors in London attended Princess Charlotte when she went into labor at full-term gestation in 1817. The labor, however, progressed very slowly. Even after 50 hours of powerful uterine contractions, the fetus wedged within the birth canal had not moved. Finally, the doctors announced to the tearful nation that a male child weighing eleven pounds had been delivered, but the future king was stillborn.

Soon the tragedy compounded: seven hours after delivering her stillborn son, Princess Charlotte died from postpartum hemorrhage.

The national grief that followed can only be compared to the collective melancholy and trauma that followed the assassination of President Lincoln in 1863, and of John F. Kennedy one hundred years later.

The doctors who attended Princess Charlotte were severely criticized. There were innuendoes and accusations, even though there was no clear evidence of malpractice. Unable to overcome his shame, guilt, and public humiliation, Sir Richard Croft, one of the four doctors attending the princess' delivery, shot himself.

The untimely death of the princess left the monarchy without an heir, which led to the usual scramble for succession in British history. Pretty soon though, a shy, obscure young lady was chosen to ascend to the British throne. Victoria at 18 became the beneficiary of a royal medical tragedy.

The rest, as they say, is history.

Chicken Farming and Baby Shows

In the late 18th century, France was reeling from the effects of Revolution. At that time the infant mortality rate was record high: more than 50% of infants were dying before their first birthday. The declining birth rates and excessive infant deaths soon led to a stagnation in population growth. This, coupled with dropping rates of pregnancy, began to alarm the country's leaders.

Particularly worried was the military brass, heavily engaged in battles with Prussia. Just when they needed more than a "few good men" to serve in the military, the population had stopped growing, and the number of men enlisting was dropping. To rectify the situation, "Depopulation Commissions" were set up, which came up with innovative approaches so that young couples could bear more children to man the future French armies.

Intense campaigns were launched to improve fertility rates; services for child health were improved; and large grants were given for research in child care. The government began to shell out money for women who became pregnant and had babies—a gift for donating one's child to the army.

That, in part, was the origin of improvements in maternal and infant care programs in 19th century France. Collectively, the plans worked; more babies began to survive and the army's recruitment increased.

In 1878, Stéphane Tarnier (1828-1897), a renowned Parisian obstetrician, was visiting the Paris zoo, and he was impressed by the poultry section, which was warm inside, despite cool weather that day. Recollecting how premature babies in his ward died from severe hypothermia, Tarnier wondered if one could construct devices similar to those that kept the poultry warm—a sort of "brooding hen" boxes—the premature babies in his unit might survive brutally cold winters.

Upon his suggestion, an engineer from the zoo devised an ingenious contraption—the "incubator." The first baby incubator was a double-walled metallic box under which a water chamber was heated with alcohol lamps, and hot water in turn heated air. As hot air circulated between the walls of the incubator, the air inside the incubator, too, was being heated. The "cage" was large enough to hold two premature babies. The box was named simply *couveuse* (incubator), which made its debut at the Paris Maternity Hospital in 1880.

To their delight, Dr. Tarnier and his students, especially Dr. Piérre Budin (1846-1907), began noting markedly improved survival of premature babies with the use of incubators. Budin wrote many books and articles about these advances, and became an internationally acclaimed expert on the care of premature babies.

Budin wanted to make the technology known to the rest of the world. He planned on exhibiting the Paris incubators at the 1896 Berlin Exposition. For this purpose, he recruited his assistant Dr. Martin Couney, and sent six incubators for the show.

While in Berlin, Couney thought of adding a sense of reality drama to the show: why not include "live babies" in the incubators? He borrowed six premature infants from a local maternity hospital and exhibited them inside the six incubators he had brought from Paris. Couney gave his premature baby show a catchy name— *kinderbrutanstalt* (child hatchery), igniting the imagination of the public thirsty for sensational scientific breakthroughs.

The show was a wild success. Couney charged one German mark per entry to see the "child hatchery." This show outdrew the Congo

Village, sky riders, and Tyrolean Yodelers. Thousands of Germans delighted in seeing the marvel of modern science in which human infants could be "incubated" inside heated cages, similar to chicks inside hatcheries.

Fortunately for Couney, all six premature infants survived, probably because he had chosen five-day-old, healthy babies who were perhaps not so premature after all.

Spurred by the Berlin success, Couney convinced Budin to continue such exhibitions and took the incubators to Great Britain's 1897 Victorian Era Exhibition.

But no self-respecting Londoner would allow premature *English* infants to be placed inside *French* incubators. So, Couney requested his chief to give him French babies. Budin complied, and as retold later, "a bunch of premature infants" were transported across the English Channel in wash baskets warmed with hot-water bottles and pillows. They became objects of exhibition in London, which again became a sensational show.

Soon there were copycats. Within eight months of the show at London's Earl Court, the incubator craze had spread all over Europe. The shows had incubator-infants breathing cigar smoke and the exhaled breath of thousands of visitors. Live leopards were kept in cages next to incubators to provide dramatic contrasts. The show owners often brought term infants claiming that they were surviving premature babies.

The medical journal *Lancet* complained: "Is it in keeping with the dignity of science that incubators and living babies should be exhibited amidst the aunt-sallies, merry-go-rounds, five-legged mules, wild animals, clowns, and penny peep-shows, along with the glare and noise of a vulgar fair?"

But such shows went on.

In 1898, Couney sailed to the US and organized the first incubator show here at the Omaha Trans Mississippi Exposition. In 1903, Couney made the US his home and began a spate of premature baby shows that lasted for nearly 40 years. He took the shows to state fairs, traveling circuses, and science expositions all over the

US. His admission charges ranged from 25 cents to $1.00. He had a permanent, annual exhibit in New York's Coney Island. About 8,000 "Couney Babies" were raised in these exhibits. Academic physicians had mixed feelings about the shows and of making a spectacle of babies, but they grudgingly recognized and accepted the benefits derived from the publicity.

The last of the baby shows was held in the 1939-40 season—now the site of a Holiday Inn in Atlantic City. A bronze plaque on the wall next to the entrance to the hotel honors Couney's shows.

Incubator Baby Shows

Little on Little's Disease

Shakespeare's King Henry VI offers one of the most poignant musings on the burdens of disability and the difficult birth (owing to footling presentation) of his brother, Duke of Gloucester, who later became Richard III.

Henry tells the duke, who was supposedly born premature, "Thy mother felt more than a mother's pain, yet brought forth less than a mother's hope."

In Shakespeare's other play about the same Duke of Gloucester, *The Tragedy of Richard III*, the protagonist laments:

> "...I, that am not shaped for sportive tricks ...I, that am curtailed of this fair proportion, cheated of feature by dissembling nature, deformed, unfinished... [and blames the reason for being so because he was] sent before my time into this breathing world scarce half made up..."

Was the man who would become King Richard III really sent before his time? Was that why he had a deformed and unfinished frame which prevented him from playing sports and made dogs bark at him?

We cannot be sure—but we are sure that it was Shakespeare

who first described an association between preterm birth and later "deformities," albeit in a fictional biography.

Some 250 years later, another Londoner made a similar association—this time with facts and figures and illustrations, and a presentation in front of a stunned group of obstetricians sitting in disbelief.

The guest speaker at the Obstetric Society of London's October 2, 1861 meeting was a London orthopedic surgeon, not the usual obstetrician.

William John Little presented a talk that had a long title, but its contents were far from boring. In the talk titled, *On the Influence of Abnormal Parturition, Difficult Labours, Premature Birth, and Asphyxia Neonatorum, on the Mental and Physical Condition of the Child, Especially in Relation to Deformities,* Little presented case histories of a large series of children with physical and intellectual impairments he had seen in his practice.

The children had suffered during birth. Their mothers' labor and deliveries were highly complicated, and included pulling the babies with forceps, or manually extracting an obstructed fetus. Consequently many children had not breathed properly for long periods after birth. Some had seizures during the first few days after birth. Many suffered from skull fractures. Little then proposed that all such "birth-related adverse events" included lack of oxygen, leading to "asphyxia," which caused those children to develop physical and intellectual impairments.

The presentation was a "learned bombshell," remarked an observer. The audience politely applauded Little, but disagreed with his conclusions. One of them said that Samuel Johnson "... *was born almost dead and did not cry...yet he became synonymous with intellectual grandeur.*"

Thus began a controversy about the etiology of a group of disorders which, in 1889, William Osler christened the *Cerebral Palsy,* or CP.

The root of the controversy was a disagreement about how to designate a "cause" or the etiology for a medical condition. Some scientists agreed with Little, but others argued that asphyxia—lack of oxygen and diminished circulation to the brain—was an "association," not the proximate cause of CP. They proposed that CP was probably congenital (inherited), or due to infections such as meningitis or poliomyelitis during childhood. Some even thought the cause of CP could be from teething-related illnesses.

There were understandable reasons why Little was misunderstood.

In the late 19th century, doctors did not understand clearly how the brain and the spinal cord worked. Some scientists were proposing the "neuronal doctrine of brain function," which explained that the brain worked the communications among brain cells, or the neurons, which are the brain's basic building blocks. This concept proved to be correct, but in the late 1800s, it was only a vague theory with no specific details. Therefore, doctors found it difficult to explain how asphyxia at the time of birth could lead to deformities in later childhood.

Little was trying to understand the reasons for poor outcomes in children already affected, by obtaining their birth histories—a research method now known as "retrospective," or "observational," as opposed to "prospective," in which one conducts serial examinations of children who had complicated birth histories. In retrospective studies, the relationship between a cause and a specific outcome is difficult to establish because a number of other factors not studied might contribute to the poor outcome as well. In Little's time, sophisticated methods, such as relative risk and odds ratios (statistical measures of association), and logistic regression analyses did not exist to help him wade through his observational findings.

Then there were other culprits: infections such as poliomyelitis, encephalitis, and meningitis occurred frequently, thus making it difficult to ascribe a single cause for a specific childhood disability. It is also likely that obstetricians disliked that an orthopedic surgeon was pointing out their errors.

Recently, neurologists have reassessed all the cases Little presented to the Obstetric Society in 1861. In 90% of his cases, Little's diagnosis of CP was accurate.

After his talk and during discussion, Little quoted Shakespeare's

King Richard's soliloquy and added that even Sir Thomas Moor said that Richard was born prematurely with "feet forward," and had teeth at birth; both were "bad omens," which forecasted Richard's deformity of the frame and cruel character.

Little concluded that King Richard III was an historical example of a man with CP who was born before his time. The truth of this assumption remains unsettled.

Tempkin and Soranus's Gynecology

OwseiTemkin died in 2002, a few weeks before his 100th birthday. A preeminent medical historian and a founder of the medical history specialty in the US, Temkin left behind an extraordinary array of scholarly works, and a large group of admiring students around the world.

Temkin was born in 1902 in Minsk, Russia. To avoid religious persecution, his family fled to Leipzig in 1905, where Owsei graduated in medicine in 1927. He studied medical history and took up a teaching position under the Swiss medical historian Henri Sigerist. In 1932, Sigerist moved to Baltimore and became the director of the first US Institute of the History of Medicine at Johns Hopkins University. Temkin came with him. In 1958, Temkin took over the position of the third director of the Medical History Institute at Hopkins.

Temkin edited the Bulletin of the History of Medicine for many years. Chief among his books were The Falling Sickness—A History of Epilepsy from the Greeks to the Beginning of Modern Neurology; Galenism—Rise and Fall of a Medical Philosophy; and The Double Face of Janus and Other Essays in the History of Medicine. At 89, he wrote Hippocrates in the World of Pagans and Christians. Ten years later, he published On Second Thought and Other Essays in the History of Medicine.

Temkin translated into English the authentic version of Soranus's

Gynecology — the only sourcebook on perinatal and neonatal medicine from antiquity.

Soranus of Ephesus, like the modern-day Temkin, was one of the most prolific writers and analytical thinkers of his time. Despite his fame, we know very little about his life. Soranus was born in Ephesus, a city in today's Turkey, and studied medicine in Alexandria.

Like many Greeks of his day, Soranus went to Rome to practice medicine and perhaps worked in the courts of Trajan (98-117) and Hadrian (117-138). More than 20 books have been ascribed to Soranus, but none has survived in its entirety. Portions of *Gynecology* and a reconstruction of its lost manuscript are now available.

The *Gynecology* is organized into four books, and reads similar to any modern textbook of perinatal-neonatal medicine. Books I and II deal with "Things Normal," and Books III and IV, "Things Abnormal."

Speaking of midwives in Book I, Soranus declares that they must be trained in "all branches of therapy" and should remain sober at all times to ensure maintaining the secrets of their patients. He insists that a midwife be free from superstitions—an extraordinary recommendation at an age when belief in magic and cults was the norm.

In later chapters, we read about the anatomy of reproductive organs; physiology of menstruation; women's health before conception; and the anatomy and physiology of the fetus. Superb chapters follow describing the signs and symptoms of normal pregnancy; of labor and the progress of labor; normal and abnormal positions of the fetus and fetuses; management of normal and abnormal labor and delivery using medicines and surgical procedures; and the care of normal and sick newborn infants.

Soranus' clear thinking and precision in writing are best exemplified in a brief section annotated below. It is noteworthy that determining the viability of newborn infants at birth remains a critical topic even today, as it was in Soranus' time.

In a chapter entitled, *How to Recognize the Newborn that is Worth Rearing,* he notes that the midwife should examine the baby, announce its sex, and note if "…its mother has spent a period of pregnancy in good health."

The midwife needs to see that the infant cries "immediately with proper vigor" and it is:

> "…perfect in all its parts, members and sense…its ducts, namely of the ears, nose, pharynx, urethra, anus are free from obstruction… natural functions of every member are neither sluggish nor weak… the joints bend and stretch…are of due size and shape and the limbs are properly sensitive in every respect. This may be recognized from pressing the fingers against the surface of the body, for it is natural to suffer pain from everything that pricks or squeezes."

This is the quintessential criteria to assess a baby at birth—equivalent to the modern-day Apgar score; perhaps the latter should be really called a "modified Soranus score."

Soranus wrote the *Gynecology* in the 2nd century. The book had an enormous impact on obstetric and pediatric practice in Rome and Greece. Over the next 1,600 years, its influence spread all over Europe and the Middle East. European and Arabic authors not only borrowed from *Gynecology*, but also added their own concepts, and implied the new additions were also those of Soranus, to maintain the credibility of the text.

Translations and translations of translations appeared, with little regard to the original. Numerous versions of *Gynecology* texts, each purporting to be that of Soranus, appeared in Latin, French, Dutch, Arabic, and Spanish; thus by the 1800s, the original *Gynecology* had been lost in translations. The oldest surviving *Gynecology* was a corrupted, 15th century Latin manuscript in France's *BibliothèqueNationale*.

In 1830, the German physician-scholar Friedrich Dietz undertook the enormous task of reconstructing *Gynecology* back into Greek. Over the next 50 years, groups of scholars in Dutch, French, and German languages along with philologists and linguists contributed to this effort. The Greek "original" *Gynecology* was completed in

1882 by Valentine Rose, and re-edited by Johannes Ilberg, another German linguist.

OwseiTemkin used this Greek version of *Gynecology* and translated into English in 1956.

The Lady and the Score

One day in the early 1950s, a student during a lunch break asked his anesthesia professor at Columbia University in New York how to assess newborn babies. The professor was struck with an idea for a score to assess babies: she scribbled something on a scrap of paper and dashed to the delivery room to test what she had written.

The restless professor was Virginia Apgar (1909-1974), the chief of the first department of anesthesiology in the United States. Her score, the Apgar score, provided grades for infants' tone, color, pulse, respiration, and reflex activity on a three-point scale from 0-to-2 for each item. Apgar learned that between one and five minutes after birth, all healthy infants scored 7 or more; the sick infants who failed to breathe or had low heart rates had scores 3 or less, needing resuscitation.

The Apgar score made its debut in 1953, and was accepted by the scientific community for its simplicity and physiological soundness. The score became the language for asphyxia and helped doctors to initiate resuscitation at the right moment.

Dr. Apgar did not stumble on to her score serendipitously. As an obstetric anesthesiologist, she had studied the effects of anesthetic agents on unborn fetuses and had witnessed the crude, inconsistent, and unscientific methods of assessment and resuscitation of infants at birth. The score was thus a natural byproduct of her contemplative mind melding her clinical and research experiences.

Virginia Apgar's career was brilliant and remarkable. She was born in West Field, New Jersey, and excelled in studies and extracurricular activities. Within months of enrolling at Columbia University College of Physicians and Surgeons in 1929, the stock market crashed, further straining her limited resources—yet she prevailed. She played violin in the college orchestra and acted in dramas; participated in seven varsity athletic teams; wrote for the school newspaper; and maintained the college library and the anatomy labs. These commitments aside, she graduated fourth from the top of her class in 1933.

Dr. Apgar began a surgical internship but changed plans, taking up anesthesiology, since women rarely succeeded in the crowded men's world of surgery. In 1938, she became the 50th board-certified anesthesiologist in the US and the first woman professor at the prestigious Columbia-Presbyterian Medical Center.

Dr. Apgar was not done: after two decades of work and research as anesthesiologist, she left Columbia to study public health at Johns Hopkins University, earning a master's degree in 1959. She then joined the March of Dimes Foundation as the director of the Birth Defects Division. She traveled, lectured, and taught about birth defects. With vigor, charisma, and enthusiasm, she helped "lift birth defects from a secret closet and put them on the map."

Dr. Apgar was a passionate teacher, consummate "learner," and a patient advocate par excellence. Her brisk walk, fast talk, fast driving, and her musical skills were equally legendary. At 50 she took flying lesions and hoped one day she would learn how to fly an airplane under the George Washington Bridge in New York. She played and made stringed instruments. There is a remarkable story of how one night, she and a colleague of hers made their way into Columbia-Presbyterian Hospital and stole a maple shelf from a phone booth, with which she later made a splendid violin.

Dr. Apgar died following a brief illness in 1974. In the fall of 1994, the US Postal Service issued the 20-cent Apgar stamp, immortalizing this remarkable American. On that day, her students and colleagues played her favorite chamber music on instruments she had handcrafted. The Apgar Quartet now performs at Columbia regularly.

In 1962, Dr. Joseph Butterfield, a former student of Apgar, proposed that "APGAR" also serves as mnemonic: "A" = appearance; "P" = pulse; "G" = facial grimace; "A" = active muscle tone; and "R" = respiration. With millions of newborn babies given "Apgars" each day in almost all modern hospitals, "APGAR" may be the most frequently uttered personal name in the world.

Virginia Apgar

Julius Caesar and the Caesarean Section

"How was Julius Caesar born?"

Most medical students assume the question to be tricky, and will give a seemingly obvious answer, "Of course, by C-section."

The question, however, was not tricky and the answer was certainly wrong.

Despite numerous artworks depicting his exit via his mother's abdomen, and folklores and anecdotes repeating the same story, Julius Caesar was probably delivered through the normal route.

Why are historians so certain?

The word "Caesar" was generic, referring to all Roman kings. The prefix "cesarean" thus meant only a royal connection—not necessarily to Julius Caesar. Moreover, since the procedure was most often fatal to the mother until the early 17th century, it was always being carried out on women who had died during childbirth. Born in about 100 BC, Julius Caesar's mother was alive, since she received letters from her son regularly during his many battle campaigns. Therefore, most historians conclude that she had not undergone the highly risky surgery to deliver her son.

So, how did the phrase "cesarean section" originate? Two equally tenable propositions have been made to explain the origin of this phrase.

One view holds that the procedure got its name from a decree by Emperor Numa Pompilius (715-727 BC), requiring the removal of fetuses from the wombs of women dying during childbirth so that the mother and her stillborn could be buried in separate graves.

This law was initially called *LexRegia* or the Royal Law. Other Roman emperors, or Caesars, also issued similar laws; thus the law acquired a generic meaning, lex *Caesarea*, or "Caesar's Law." As later recorded in Justinian's book *Degesta*, the law reads, in part:

> "The lex regia (Royal Law) forbids the burial of a pregnant woman before the young has been excised: he who does otherwise clearly causes promise of life to perish with the mother."

The phrase "promise of life" implies that at least in some instances, babies so delivered might have survived, and therefore, on rare occasions doctors carried out cesarean sections to save a potentially viable fetus.

Saving the unborn had an important pragmatic reason: the husband had a claim to his dead wife's wealth and assets only if the child she delivered was a son *and* was alive—not a stillborn son or a living daughter. This applied even if the infant died later. Therefore, besides the burial-related ritualistic reasons, the potential for inheritance might have been a powerful motive behind the Roman emperors' *LexRegia,* and for its continued practice.

The second explanation for the origin of the phrase "cesarean section" is much more mundane: it originated from the Latin term *caeder,* meaning "to cut." The noun form of this became *partuscaesareus*—the "cesarean partition," or section.

Delivering a live fetus after cesarean on dead women was extremely rare. Therefore, such infants were deemed special and fortunate, even supernatural, possessing divine or demonic powers because they were "not of a woman born," but of a corpse. Several historical figures and mythological characters have been described as having been delivered via the cesarean route.

Having learned of his wife Coronis' unfaithful nature, Apollo had her slain. Later, out of remorse, he removed the unborn son Asclepius from her womb as she lay on the funeral pyre. Asclepius became the god of medicine and health, and his shaft with a single entwined serpent became the symbol of the medical profession.

In Hindu mythology, Lord Brahma, the god of the Vedas, is described as being born through the umbilicus of another god, Lord Vishnu. Sixth century stone sculptures in Indian caves show the Buddha being delivered through the right flank of his mother, Maya. Based on Talmudic writings, some experts believe that in many ancient Jewish tribes, C-sections were being carried out on living women.

In *Macbeth* Shakespeare declares that "none of woman born / Shall harm Macbeth." The prophecy holds true when Macduff tells Macbeth before slaying him that he, Macduff, "was from his mother's womb / Untimely ripp'd."

The first documented cesarean section on a living woman was performed in 1610 by Jeremius Trautmann, a surgeon from Wittenberg, Saxony, in Germany. He reports that the mother survived for 25 days.

The first documented cesarean section performed in the United States is credited to Dr. Jesse Bennett at his log cabin in Mason County, Virginia (now in West Virginia). On January 14, 1794, he operated on his wife under a heavy dose of laudanum. After swiftly delivering a healthy baby girl, he also removed both the ovaries of his wife, reportedly saying that he wished to avoid such an ordeal again. Both the mother and child survived.

Dr. Bennett did not report his own success, but made marginal notes in a personal copy of a 1801 obstetric textbook he owned.

This book is in the Richmond Academy of Medicine Inc.

Dr. Pap's Gift

A Greek doctor and his wife landed on Ellis Island on October 19, 1913. They knew no English, had no friends, but had $25 as fee for their visas and lots of hope for their future. They found work, he as a rug salesman and she as a tailor. He also played violin at restaurants and worked as a clerk in a Greek-language newspaper. Later, he got a research job in the anatomy department at Cornell Medical College, and his wife became his assistant.

Thus began the extraordinary career of Dr. George Papanicolaou (1883-1962) and his wife, Andromache (Mary), in the Land of Opportunities—in return, they gave a gift, the "Pap smear," that would save millions of lives.

George Papanicolaou was born in Kymi, a village on the island of Euboea, Greece. His father, also a doctor, served in the Greek National Assembly. George graduated with honors in medicine from Athens and completed his required military service. But realizing that his calling was biological research—not patient care—he went to Munich and earned a PhD in zoology in 1910.

On a ferryboat trip, he met and fell in love with Mary. After marriage, George served again in the Greek military, and fought in the 1912 Balkan Wars. Soon after, the couple migrated to the United States to pursue research careers, in spite of strong objections from their families.

In 1917, Dr. Papanicolaou began studying sex chromosomes in guinea pigs. He needed to obtain eggs from female animals in early stages of their division. This was technically very challenging in guinea pigs that were alive.

It seems that one night he had a dream that like in humans, guinea pigs, too, must have "menses," and the amount of the vaginal bleed would be very small. Thus he reasoned, by obtaining a smear from the animal's menses-blood he might be able to obtain unfertilized eggs normally expelled during menses.

The next morning he obtained a smear from a guinea pig vagina, in what was probably the first "Pap smear." Under the microscope the smear revealed "an impressive wealth of diverse cells." He began analyzing cellular changes as seen in such smears obtained during guinea pigs' estrous cycles.

As Dr. Papanicolaou started publishing his results, the *method* of studying the cells "shed" from the vagina became standard for testing the effects of hormones and drugs on reproductive organs. Soon Dr. Papanicolaou began collaboration with other doctors, and extended the technique to study vaginal cytology in humans.

While examining specimens from normal volunteers, Dr. Papanicolaou encountered a single specimen that contained malignant cells. To his great credit, he did not discard the specimen as an artifact. Instead, noting the moment as "the most thrilling" of his career, he began studies using the vaginal smear-specimen from women with known genital malignancies.

This process helped him to discover that cancerous cells in the vagina were being shed in those with malignancy, which led him to conclude that discovering shed cells should be one of the means for diagnosing uterine cancers.

When Dr. Papanicolaou presented his results at the third Race Betterment Conference in Battle Creek, Michigan, in 1928, cancer experts were not impressed. They held onto the prevailing view that cervical biopsies were much more definitive for the diagnosis of cancer than were the "blindly obtained smears."

After ten more years of relative obscurity, Dr. Papanicolaou's vaginal smear method began to be tested in clinical trials, establishing the usefulness of this procedure as a screening tool, since it also revealed pathological cells in pre-cancerous states. Dr. Papanicolaou published a landmark paper in 1941, and the "Pap smear" test became a standard for screening women.

How good was implementing the test in outcome of women with cervical cancer?

According to some studies, due to early detection, the Pap smear reduced the cervical cancer death rate by 70%. Among those not screened, 14 women out of 100,000 died from cancer, whereas among those undergoing routine screening, only 4 per 100,000 died—a 70% reduction in death rate. Performed properly, the error rate is under 5%. Pap smears can also detect vaginal infections, including those from herpes virus and human papillomavirus.

Dr. Papanicolaou made the short-list for the 1960 Nobel Prize in Medicine. The Greek government honored him in 1973 by issuing two postage stamps with his likeness and in 1995 by printing his picture on its 10,000-drachma banknote. In1978, the United States, too, issued a stamp in his honor.

The graduating class of Cornell Medical School takes its Hippocratic oath under a maple tree planted in his honor in front of the school. The tree is a sapling from the original Hippocratic maple from the island of Cos.

Dr. Papanicolaou's research would not have succeeded without the collaboration of his colleague, confidant, and willing guinea pig for many of his studies—Mary Papanicolaou. He called her "my wife and my victim." True to her name Andromache, literally, "a woman fighting with men," she stood by her husband and "fought" alongside.

Was Dr. Papanicolaou's great discovery serendipitous? Probably not:

"...Chance favors only the prepared mind," said Louis Pasteur.

Interpreter of Hearts

Maude Elizabeth Abbott (1869-1940) made fundamental contributions to the phenomenal growth of pediatric cardiology in the 20th century like few other contemporaries.

In 1890, Maude Abbott graduated as the valedictorian from McGill University in Montreal, majoring in arts. She was the first woman to be selected as the class president. Thus she felt assured that her lifelong dream of becoming a doctor would be possible. But her application to her "beloved McGill" was rejected because no woman had ever been admitted to the McGill School of Medicine. One professor of surgery even threatened resignation if any woman was granted admission. Other faculty supported Abbott's application and the local newspapers protested—but the university's decision prevailed.

About this time, Abbott received an offer from the newly opened University of Bishop's College in Montreal, a rival of McGill. She accepted and became the first and the only female medical student at Bishop's.

Abbott was no stranger to adversity. She was born Maude Babin in St. Andrews East, Quebec. Her father deserted the family soon after she was born, and her mother died of tuberculosis when Maude was not even one year old. A maternal grandmother adopted Maude and her sister Alice, and gave them her own family name, "Abbott."

Maude and Alice were tutored at home. Girls did not attend high school in those days, but since the first high school for girls had just opened in Montreal, Maude was encouraged to enroll. In 1885, she graduated, winning a scholarship to study at McGill.

Abbott graduated from the Bishop's medical school in 1894, winning the Chancellor's Prize and the Senior Anatomy Medal. She then went to Europe, taking Alice along with her. The trip— the mandatory "pilgrimage" for new medical graduates seeking postgraduate medical training—led her to meet some of the brightest minds in clinical medicine and pathology in Europe. While in Vienna, however, tragedy struck: Alice developed diphtheria and sustained permanent brain damage. Maude took care of her sister for the next 40 years.

Back in Montreal, Abbott opened a medical practice in 1897. Dr. Charles Martin, one of her teachers from medical school, advised her to do research at Montreal's Royal Victoria Hospital. Abbott's first project was to study heart murmurs. Thrilled by the chance to do research, Abbott cultivated a lifelong passion to study children's heart diseases.

In 1898, Abbott took a job as the assistant curator at McGill University's Medical Museum, where she started examining literally thousands of pathology specimens collected over several decades. She visited other museums in Canada and the United States to learn more about medical museums.

On a trip to Washington, DC, Abbott met Dr. William Osler, professor at Johns Hopkins in Baltimore, who was formerly at McGill. Osler was thrilled to discover Abbott's interest in pathological anatomy—a subject dear to Osler, too. He advised her to study the subject systematically and invited her to write a chapter for his textbook. When she asked how she should approach the subject, he told her, "statistically."

With a rigorous approach to the study of the specimens at the Medical Museum, Abbott became an expert on the pathology of the developmental abnormalities of the heart. In 1901, she became the curator to the museum; by 1904, McGill medical students were required to take a museum rotation and a course under Dr. Abbott.

Thus, Abbott, who was denied admission at McGill, became a faculty member there. A fire at the museum in 1907 destroyed hundreds of pathology specimens, but Abbott's catalogues were saved. She then took on Osler's invitation and wrote a chapter on congenital heart disease for his 1908 edition of *Modern Medicine.* Osler was delighted, as it was a definitive chapter on cardiac malformations in children, which had not been studied systematically until then.

Until the late 19th century, it had been thought that heart lesions could not be treated surgically. By explaining the anatomical defect and defining the hemodynamic consequences of the abnormalities, Abbott's work provided a firm, scientific basis for the understanding of cardiac disorders. Using this knowledge, future surgeons developed life-saving operative procedures.

Abbott described the anatomy of a disorder now known as *the transposition of great vessels,* which she described as "a condition in which the position of the large blood vessels leading out of the heart had been reversed." It was the first perfectly accurate description of a disorder that had been enigmatic until then.

Abbott's collected works appeared in the superbly illustrated classic, *An Atlas of Congenital Cardiac Disease.* In 1907, Abbott founded the International Association of Medical Museums and served as the secretary and editor of its journal until 1938. Through her writings on science and on the history of medicine, she became a internationally acclaimed pathologist. Yet, McGill University did not offer her a full professorship.

Abbott continued her meager clinical practice, looked after Alice, and did research at McGill's Museum. In1923, she took a position at Women's Medical College of Pennsylvania in Philadelphia, and two years later returned to McGill as Assistant Professor of Medical Research. However, at 67—the mandatory retirement age—she was forced to retire.

Many professional organizations honored Abbott. She was the first Canadian to be named a member of the California Heart

Association and the New York Academy of Medicine. After her retirement, the Carnegie Foundation gave her a grant for her to write all she had learned about heart disease. Unfortunately, this work was never completed. Abbott suffered a stroke in July 1940, and died in September of that year.

In 1999, Canada Post commemorated Dr. Abbott by issuing a stamp in its Millennium Collection Medical Innovator set.

Blue Babies

In the 1930s, accurately diagnosing children's heart diseases was almost impossible. Ultrasound, or echocardiography, had not been invented. Cardiac catheterization was almost never carried out in children, and children's EKG tracings looked so different from those of adults, even most cardiologists found them impossible to interpret. Chest X-rays were the only tool for diagnosis, which provided too little details about the anatomy of the heart's interior. So, when an accurate diagnosis of a heart defect was really made, at best it was an educated guess, and at worst, confirmed at an autopsy.

Into such an underdeveloped world of medical specialty entered Dr. Helen Taussig (1898-1986) in 1927. But, for the young lady who was to earn a reputation as the "First Lady" of pediatric cardiology in the United States, the career trail was far from smooth.

Her father was a leading economist in Cambridge, Massachusetts. She wanted to study medicine at Harvard Medical School, but women were not allowed at Harvard. So, she decided to study public health, and applied for a seat at Harvard's School of Public Health. During the interview, the dean told her that she may study there, but would not be granted a degree because she was a woman.

She asked him, "Who is going to be such a fool to study [for four years] without getting a degree?" The dean said, "No one, I hope," to which Taussig said, "Dr. Rosenau, I will not be the first to disappoint you," and enrolled there anyway.

Discrimination because of her gender continued at Harvard. In one instance, during a histology class at Harvard, she was made to sit in a remote corner, "so as not to contaminate Harvard's [male] students," remarked a teacher. Frustrated, Taussig requested a transfer to Boston University to pursue studies in public health. She soon obtained an entry at the Johns Hopkins Medical School, moved to Baltimore, and graduated in 1927.

But for Taussig, even at Hopkins things did not go smoothly. She was denied an internal medicine residency because she was a woman. She then chose pediatrics as her field of specialty—for the children of the world, this turned out to be a lucky move.

While training as a pediatrician she focused on heart diseases in children. In 1930, Hopkins allowed her to start and run a new pediatric cardiology clinic—the first of its kind in the United States.

Taussig is remembered for having developed treatment for a common heart condition and saved millions of babies from dying early.

In 1880, Etienne-Louis Fallot of France described the anatomical features of a common heart ailment affecting children. Four anatomic abnormalities within the heart were evident in this condition: a blockage at the pulmonary artery—a major vessel taking blood from the right ventricle of the heart to the lungs; a hole in the wall separating the left and the right ventricles; a displacement of the aorta to the right; and a severely thickened wall of the right ventricle. Because of four abnormal heart features, the disease was called the *Tetralogy of Fallot*, abbreviated as TET. Its victims were infants and toddlers who were always blue, would get tired easily, and become chronically debilitated. Most "blue babies" died before adolescence.

Taussig developed a systematic method to study the blue babies. She charted their clinical signs and symptoms and how they changed as they grew. She studied their hearts under fluoroscopy X-rays. She attended the autopsies of TET victims and gradually figured out what killed them. It was the lack of oxygen that killed them—not the structural abnormalities in the hearts.

The basis for her conclusions was that during intrauterine life, the placenta provides oxygen to the fetus—not the lungs. Therefore, lungs are inactive, receiving little blood from the right side of the heart. About 80% of the blood from the right ventricle is diverted or "shunted" away from the lungs to the left side through a connecting tube called the *ductusarteriosus* (DA). This tube is hooked up to the aorta from the pulmonary artery, and in healthy children it closes off spontaneously within a few days after birth.

In some babies who have otherwise normal heart, this tube may remain open, leading to a condition called *patent ductus arteriousus* or PDA. A large PDA can lead to symptoms of heart failure. Such babies need heart surgery to tie off their PDAs. The operation for the PDA, called the PDA ligation, was relatively simple. Therefore, it was the only procedure surgeons dared to carry out in children with any heart ailment in the 1930s.

But Taussig figured out that some TET babies who also had a PDA did not become ill—in fact, the PDA seemed to help them. TET babies with PDA were less blue, had fewer symptoms of fatigue and exhaustion, and lived longer. By contrast, TET babies in whom PDA closed spontaneously soon after birth died within a few weeks. It was clear that TET babies benefited from a PDA, which increased blood supply to the lungs, increased oxygen supply to the body, and prevented early death.

Then Taussig wondered, why not *create* an artificial PDA in TET babies? She made a bold proposal to the young thoracic surgeon at Hopkins, Alfred Blalock: "If you can close a PDA, why can't you 'create' one between the aorta and the pulmonary artery to increase blood flow to the lungs?"

Pediatric Cardiologist Dr. Helen Taussig

Blalock took up the challenge, and conducted animal experiments to perfect the technique of making vascular shunts. In November 1944, he performed the first shunt surgery on a blue TET baby, while Taussig monitored the patient. After a stormy postoperative course, the baby recovered. After two more such procedures, they presented their results to the scientific community.

Over the next four years, Johns Hopkins University Hospital became a renowned center for blue baby surgeries. Each year, the duo performed over 100 "Blalock-Taussig shunts."

Taussig traveled around the globe teaching other surgeons the procedure. BT shunting remained the initial palliative procedure for severe TET and other similar cyanotic heart conditions until very recently.

Now, because of improved surgical techniques, anesthesia, and advances in intensive care for critically ill newborn infants, most cardiac surgeons prefer to perform total reconstruction of the heart abnormalities in infants who manifest symptoms of TET and other conditions that lead to "blue babies."

Taussig's career depended upon listening to heart sounds—but, sadly at age 40 she began to lose hearing. Since childhood she had been dyslexic. Yet, neither of these disabilities curtailed her work; her unparalleled energy, brilliant intellect, and incredible kindness earned her admiration from her students, colleagues, and patients the world over. Taussig was a legend.

Some Famous "High-Risk" babies

Saints and rivers have mysterious origins goes an Indian proverb. Perhaps: but, the origins of some celebrities at birth, though not mysterious, might well have been catastrophic.

Some examples: Although for some of these stories compelling records of accuracy do not exist, one can use them to ease anxieties of parents burdened by the birth of their high-risk babies. Parents can feel that not all is lost, despite the rough starts of their babies. These stories are fun to read, too.

Religious Leaders:

- Moses: According to Jewish tradition Moses was born "six months and one day" after he was conceived. This allowed him to be hidden for three months from Pharaoh's soldiers searching for the liberator of Jews.

Historical personalities and fictional characters:

- Duke of Gloucester (later Richard III, 1452–1485): He was probably born prematurely, with feet first, and might have developed cerebral palsy. It is unclear if he had hemiplegia—paralysis of one side of the body.

- Macduff (Scottish nobleman in Shakespeare's *Macbeth*): Delivered by cesarean section after his mother's death, he was "not of a woman born," but born of a corpse; thus he was destined to kill Macbeth.

Artists and writers

- Jonathan Swift (1667–1745): Some historians claim that he was born preterm.

- LicetusFortunio (1577–1657): "A fetus no more than five and one-half inches" at birth. His father, a doctor, raised him in an oven, "similar to the chicken hatching method used in Egypt." Fortunio became a scholar and wrote 80 books.

- Pablo Picasso (1881–1973): At birth, he was blue and did not breathe. Thinking him to be dead, the midwives left him on the table. His uncle, Don Salvador, a practicing doctor, resuscitated him by providing mouth-to-mouth breathing.

- Voltaire (1694–1778): He was born preterm and appeared asphyxiated. The "puny little boy" was not expected to live and thus was baptized hurriedly. He was raised in an attic to maintain his body temperature.

- Samuel Johnson (1709–1784): He was described as "huge at birth," though no one weighed him. He was also "strangely inert" and required slapping and shaking to help him breathe and cry. After several minutes, "with persuasion" he made a few whimpers.

- Johann Wolfgang von Goethe (1749–1823): After three days of labor, his mother delivered him. He was "lifeless and miserable" and thought to be stillborn.

- Anna Pavlova (1882–1931): The famous Russian ballerina was "premature, so puny and weak" that she was wrapped in cotton wool for three months.

- Thomas Hardy (1840–1928): At birth, he was thrown aside as dead. "A good slapping" from the midwife revived him.

- Sidney Poitier (born 1927): Being three months premature, he was so small that his father "could place him in a shoebox." His grandmother said that even though a small and premature baby, he would "walk with the kings." He did. As the ambassador to Japan from the Bahamas in the 1990s, he walked with the King of Japan. (Quoted by Sidney Poitier in a television show, "Biography," CNN, Spring, 2000.)

Scientists

- Johannes Kepler (1571–1630): He was a "seven-month" baby; estimated IQ, 161.

- Christopher Wren (1632–1723): An historian, Dr. Thomas Cone Jr., wrote that Wren was born prematurely. Details are not known.

- Isaac Newton (1642–1727): At birth, he was thought to be "as good as dead." He was such a "tiny mite" that he could be placed in a quart mug.

Politicians

- Franklin D. Roosevelt (1882–1945): Weighed ten pounds at birth. He was "blue and limp with a deathlike respiratory standstill," said the midwife. This was probably from the effects of chloroform given to his mother, Sara Roosevelt, during labor.

- Winston Churchill (1874–1965): His early birth "upset the ball." Later a duchess remarked that the baby had such a lusty, "earth-shaking cry," she had never heard from a newborn before. Recent historians doubt about his premature birth.

No More Hot Oil Treatment

Doctors in the 16th century treated gunshot wounds in their patients by pouring boiling oil over the raw wounds. They justified this most cruel treatment with a simple explanation: Scalding oil "neutralized" gun powders which would otherwise poison and kill the patient. The treatment, however, caused horrible pain, destroyed living tissues, and initiated gangrene, often requiring amputation of limbs. Death was also not infrequent.

Thankfully, an unexpected scarcity of oil led a humane surgeon from France to discover the truth and change the practice.

The doctor was Ambroise Paré (1510-90). A contemporary of the anatomist Andreas Vesalius, Paré was born in a small village near Laval, France. Although Paré witnessed many advances in medical treatment during his youth, the surgical treatments remained primitive. Just as trained physicians performed surgical operations, barbers, too conducted surgical procedures. They cauterized patients' bleeding sites, fixed broken bones, and sometimes did major operations. Physicians looked down upon surgeons as barbarians.

Paré came from a family of barber-surgeons: His brother and brother-in-law were barber-surgeons, too. At 15, Paré followed the family tradition, worked as apprentice under another surgeon, and later worked on his own in the only public hospital in Paris. But Paré received his true education at the battlefields of Europe, where he served as a military surgeon.

Like his peers, Paré, too, began treating gunshot wounds with boiling oil. During the 1537 Turin battle, he ran out of oil, forcing the anguished, but quick-thinking surgeon to improvise. He dressed the wounds of his patients using egg yolks, rose oil, and turpentine. Later he would write:

"That night, I could not sleep at my ease, fearing [that due to] my lack of cauterization I should find the wounded…dead or impoisoned."

That was not to be: the next morning Dr. Paré was overjoyed to see that all soldiers who had received the soothing bandage were better, and those treated with scalding oil were in agony. Then and there Paré resolved *"never so cruelly to burn [the] poor wounded."*

In 1552, Paré reintroduced the practice of tying off arteries with ligatures during amputations instead of cauterizing them with hot iron rods. He also invented many surgical instruments, including artificial limbs, and special types of probes and forceps.

Paré's surgical skills and teaching brought him much fame. He was elected to the official college of surgeons. He wrote 25 books on medicine and surgery, and served as surgeon to four French kings. Charles IX, a Catholic king, held Paré in such great esteem that he sheltered the Protestant surgeon in the palace during the 1572 massacre of Huguenots on St. Bartholomew's Day in Paris.

Paré understood the deeper significance of the art of healing. He practiced and taught that besides being good at surgery, a doctor must be kind, gentle, and compassionate. Historians call Paré the Father of Surgery, for he had rescued the reputation of surgery that had been looked upon as barbaric.

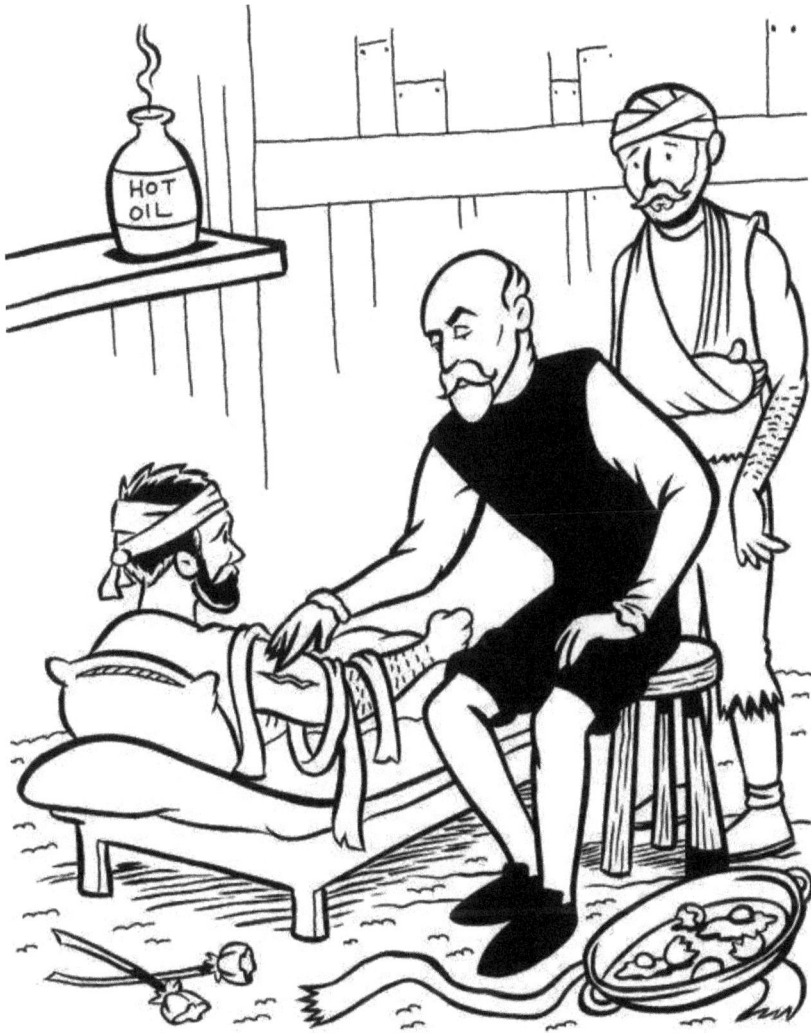

AmbroiseParé

Self-Experimentation
in the Extreme

A 25-year-old surgical resident in Germany in 1929 performed one of the most daring experimentations in history—he inserted a catheter 65 centimeters long into his own heart, and took an x-ray to show that catheterization of a living human heart is not dangerous. Seventy years later, Renate Forssmann-Falck recounted her father's experiment that day: "Forssmann, headstrong and intense believed that the experiment was absolutely safe and, grandiloquently, performed the catheterization on himself..."

A series of historical developments spanning many centuries were foundations for Forssmann's self-experimentation.

William Harvey published in 1628, *De Motu Cordis* (On the Motion of Heart and Blood)—a landmark in medicine. Harvey was the first to explain, completely and accurately, how the blood in the body circulated. It flowed in a circular fashion and it was "the same blood" that went around in the veins and arteries.

The next major development was in 1727 when Stephen Hales, an ordained priest from Kent, measured blood pressure by inserting long glass tubes into the arteries of dying horses. He concluded that the height of the blood column that rose in the tube was the "pressure of the blood"—although he, nor anyone else during his time, had not a clue about the clinical significance of blood pressure.

More than a century later, Claude Bernard (1813-1878) measured the blood temperature by inserting a mercury thermometer via the carotid artery of a horse.

By the turn of the 19th century, scientists were adept at putting catheters into the hearts of experimental animals, but they did not dare to the same in living humans. Doctors lacked safe, flexible catheters for maneuvering the twists and turns along the way from peripheral veins into the hearts; they lacked drugs to prevent formation of clots around the inserted tubes. But the most important reason for staying away from the catheterization of human hearts was one of fear—what if the patient died during catheterization?

After graduating from medical school in 1928, Forssmann took a surgery residency at a teaching hospital near Berlin, hoping to change to the specialty of medicine. During medical school, Forssmann had read about experiments by Harvey, Hales, Bernard, and others who had used catheters to measure pressures in the heart chambers of animals.

Forssmann wondered if it could be possible to introduce medications directly into the hearts rather than pushing them from peripheral veins, since drugs such as chemotherapeutic agents were too toxic to blood vessels.

But how should one insert catheters into hearts?

After threading rubber tubes into the veins of experimental animals, he then tried it on cadavers. Once he gained this experience, he approached his chief, Dr. Richard Schneider, Chair of Surgery, asking to perform such procedures in hospitalized patients; Schneider objected, fearing complications.

Young and reckless Forssmann took the matter into his own hands. He befriended an unsuspecting operating-room nurse and convinced her that his experiment would be safe. In an unoccupied operating room, as the horrified nurse looked on, Forssmann dissected his own vein near his left elbow, inserted a long "well-oiled" catheter that was normally used to remove kidney stones, and threaded it towards his own heart.

With the catheter in his heart, both of them walked to the x-ray room. Forssmann looked at the mirror held by the nurse that reflected the fluoroscopic x-rays image while as he pushed the tube further into his hear. He constantly kept watching the image of the catheter tip while slowly pushing it towards his own heart. Maneuvering the catheter around the armpit was difficult. He felt pain and tingling sensations when the tube turned towards his heart. Yet, he managed to push the tube well into his heart. He then took an x-ray that showed the catheter tip in the right heart—the first x-ray of a catheter in a living human heart.

Within an hour, almost everyone in the hospital had heard of the sensational story of how Forssmann had performed cardiac catheterization on himself. But not all were happy. Furious that Forssmann had disobeyed, Dr. Schneider reprimanded Forssmann severely.

Forssmann sent a brief report of his experiment along with a copy of the x-ray showing the catheter in his heart to the medical journal *KlinWschr*, which published it in November 1929. By then Forssmann had been transferred to another hospital. His article was widely quoted in the local newspapers. Upon learning this, his current chief was furious; he instantly fired Forssmann.

Forssmann tried to remain in academia, and tried several jobs; each lasted for only very short periods. Forssmann quit academia completely, finished his surgical training, and settled in a remote border town in Germany to practice urology.

In the 1940s, scientists in the United States, Andre Cournand (1895-1988) and Dickinson Richards (1895-1973) and their colleagues explored methods to study the heart and rediscovered the forgotten paper by Forssmann. They adapted his technique to study cardiac functions in animals and in patients. Cardiologists worldwide accepted the procedure of cardiac catheterization as a major diagnostic and therapeutic advance, saving millions of lives.

The Nobel Foundation awarded its 1956 Nobel Prize in Physiology or Medicine to Cournand, Richards, and the all-but-forgotten Forssmann "for their discoveries concerning heart catheterization and pathological changes in the circulatory system."

The Forssmann story is not complete without a reference to his Nazi past.

From the early 1930s through the Second World War years, Forssmann was a member of the Nazi Party and served in the Nazi army in 1939. During this time, millions of Jews were tortured and persecuted. German doctors conducted unethical experiments on thousands of Jews—many in the Moabit Hospital (renamed Robert Koch Hospital), where Forssmann held several posts, including that of the Vice Chair of Surgery from October 1936 through July 1938. These have never been disputed.

There is no record of complicity by Forssmann on Nazi atrocities— but a scathing letter from Dr. David Seigel in the *American Journal of Cardiology* in 1997 and a reply by Forssmann's daughter in the same issue reveal a complex story.

His daughter wrote that her father was both a victim of his own tortured personality as well as that of the political circumstances surrounding his life and times. "Speaking as his daughter... he took responsibility for his Nazi past... he did not conceal his knowledge of Nazi persecutions..." she wrote, concluding, "A historian, not a daughter, must make sense of the incomprehensible."

Hot Brains

Patients suffering from syphilis dreaded one of its most severe complications: syphilitic brain infection, or *general paresis of the insane* (GPI). Until the advent of penicillin in the 1940s, there was no effective treatment for syphilis, much less for GPI.

In the 1890s, a Viennese doctor introduced "malaria therapy" to cure GPI—an adventure that reads like science fiction rather than medical practice. The central figure in this story was Julius Wagner-Jauregg (1857-1940), a psychiatrist specializing in experimental pathology.

After medical graduation, Wagner-Jauregg wanted to specialize in internal medicine, but he could not secure a residency. He then decided to study psychiatry, because as he noted later, [his choice] "harmed neither myself, nor psychiatry."

Wagner-Jauregg had a busy psychiatry practice. While treating his patients who had both psychiatric and neurological symptoms, Wagner-Jauregg had discovered a curious association.

Some of his psychiatric patients reported that their symptoms had improved after they recovered from bouts of fevers. Wagner-Jauregg then systematically documented the signs and symptoms of 30 psychiatric patients who had suffered typhoid, malaria, smallpox, scarlet fever, and erysipelas and confirmed that in a small percentage, their psychiatric symptoms indeed got better when they had recovered from fevers, irrespective of what caused the fever.

An old dictum in folklore medicine was that it was possible to treat "one disease with another disease." Extending this concept, Wagner-Jauregg wondered if fevers improved psychiatric illnesses, at least some psychiatric conditions that might be due to specific abnormalities in the brain (rather than in the "mind"). These were called "organic" causes. Therefore, he thought such conditions might be cured by "organic" remedies—not by psychotherapy.

He faced the problem of how to induce fevers in his psychiatric patients.

Wagner-Jauregg first injected tuberculin in his psychiatric patients to induce fever. Robert Koch had developed tuberculin, hoping it would be a vaccine against tuberculosis (which it was not to be). However, in Wagner-Jauregg's psychiatric patients, tuberculin neither caused fevers regularly nor did it abate their symptoms consistently.

Wagner-Jauregg kept pondering about the manner in which he could induce fever. Then, an opportunity arose in 1917: a soldier with malaria was admitted to his clinic at the same time when many patients suffering from GPI were also in the ward. Wagner-Jauregg was struck with a brilliant idea: why not induce malaria in his psychiatric patients? He drew out some blood from the soldier and injected small amounts of the same into the arms of four patients with GPI.

All four developed malaria. He then drew out *their* blood samples and injected the blood into five other GPI patients, all of whom developed malaria, too. He did not cross-match blood samples to check blood groups. Even though blood groups had been discovered in the early 1900s, no one paid attention to cross-matching prior to transfusions. In the nine patients who received malaria blood, Wagner-Jauregg did not treat their malaria with quinine until each had gone through at least seven to ten bouts of fever.

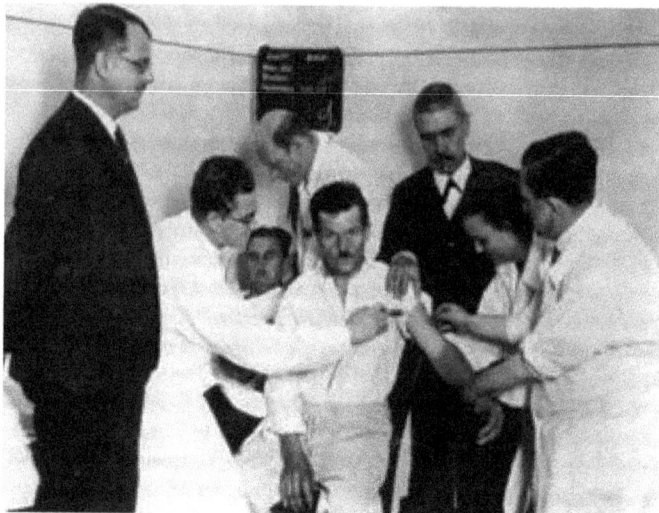

Wagner-Jauregg (suited, standing behind the patient), supervising injection of a malaria-patient's blood to a psychiatric patient.

How did his nine patients fare?

Three left his clinic alive and reportedly remained well for one year. Three improved initially but worsened later. The psychotic symptoms were unaltered in two, one of whom developed a severe "paralytic melancholy." One other died from fever.

This led Wagner-Jauregg to conclude that six out of nine had recovered. They were "better than expected," he later claimed. He published his results in 1918.

Within five years, Wagner-Jauregg's "malaria fever-cure" began to give hope to desperate psychotic patients suffering from GPI. Doctors in many countries began fever therapy for mentally ill patients. They induced fevers using all sorts of techniques on all sorts of psychiatric disorders, including schizophrenia. Some patients recovered, but most did not. Death was not uncommon as a result of malaria treatment.

By the mid 1920s, Wagner-Jauregg was world famous. In 1927, he was awarded the Nobel Prize in Medicine or Physiology—the first psychiatrist to win this prize.

Scientific Monthly said that "...the whole...world should join his patients and students in their congratulations." Even Sigmund Freud offered his congratulations.

Wagner-Jauregg's malaria therapy for GPI appears gruesome, unethical, and unscientific. But, during Wagner-Jauregg's time, this risky therapy *was* the best one could offer for psychiatric patients. Until the development of penicillin, a diagnosis of GPI was a death sentence. Even for other psychiatric conditions, there were no drugs until after the 1960s.

Some historians credit Wagner-Jauregg as a pioneer for developing "physical" (somatic) cures for psychological illnesses rather than psychotherapy —an idea that was accepted only a century later. Others criticize him as a "dyed-to-the-core" anti-Semite, who experimented on poor Jewish patients without their consent in the name of science. After the Second World War, revelations of such atrocious experiments conducted by German scientists in part led international agencies and organizations, scientists and ethicists, and other political and professional groups to develop rules and regulations, such as the Nuremberg Code of International Ethics, as foundations for the ethical conduct of research on humans, also requiring prior informed consent.

Martin's Fistula and Marshall's Ulcer

For centuries, doctors faced difficulties in understanding the mysterious workings of the digestive system, and in diagnosing the ailments affecting the stomach and the intestine. Some researchers adapted unconventional methods to probe the gut.

Frederick II of Prussia (1712-1786) designed an experiment to test if exercise helped digestion of food. He fed two hungry prisoners lots of food; afterward one was ordered to exercise vigorously and the other to rest. After the first prisoner had finished his workout, the king ordered both prisoners executed at the same time. Then he had their bowels removed to examine which of them had digested the meal better.

Great kings needed no Institutional Review Board to approve their human experiments, nor did they require informed consent from anyone.

About a century later, Dr. William Beaumont (1785-1853), a Connecticut surgeon, adapted a more humane method to study gastric physiology.

On June 6, 1822, Alexis St. Martin, a French Canadian, wounded himself in the chest and abdomen following an accidental gunshot. He was brought to Beaumont's care. For several months, Beaumont nursed him to recovery. The abdominal wound healed, but a hole remained in St. Martin's stomach, which connected to his skin through a fistula or a track.

The army had declared St. Martin a "common pauper," and they ordered him to be sent back to Canada. But Beaumont felt that such a move would be fatal to St. Martin, since his stomach hole had not healed completely. Deciding to keep his patient, Beaumont hired St. Martin as a servant.

From 1825 to 1833, Beaumont studied Martin's stomach physiology, benefitting from the unhealed hole in his patient's stomach. He collected Alexis's gastric juice during various routines, documented the chemistry and physiology of the stomach, and published the results in *Experiments and Observations on the Gastric Juice and the Physiology of Digestion.*

Fast forward to 1982:

Barry Marshall (b 1951), a senior medical resident at the Royal Perth Hospital in Australia, was looking for a research project to fulfill his training requirement. Three years earlier, J. Robin Warren (b 1937) the hospital pathologist, discovered small, S-shaped bacilli in the biopsies of stomachs from 135 patients who had gastritis—inflammation of the stomach. Warren asked Marshall to grow the organisms in small discs (a process called culture) to study the microbes further, and to correlate patients' symptoms with the biopsy findings.

Marshall worked in a corner of the laboratory for his culture studies, and called it the "fecal lab." He was soon frustrated because even though he saw the tiny microbes from biopsy specimen under the microscopes, he could not culture them even at 48 hours—the duration by which most disease-causing microbes could be cultured. As per the protocol which called for discarding the specimen after 48 hours of negative culture and declaring the patient infection free, the laboratory technicians would discard Marshall's culture plates after 48 hours of no growth.

But, on the 1982 Easter weekend, a technician, in a hurry to go home, forgot to discard several of Marshall's culture plates on which biopsy samples from gastritis patients had been inoculated for culture. Returning for work after the long-weekend break, Marshall saw that in one of those plates, plenty of microbial colonies were growing. Under the microscope he saw S-shaped bacteria that were noted previously in the biopsy specimen. Warren and Marshall published these findings in June 1983 *Lancet*, in which they made a bold, tongue-in-cheek prediction: those organisms (they called it *Campylobacter-like* microbes) *might* be the cause of stomach ulcers and possibly of stomach cancers.

Although their findings were treated as a breakthrough, most scientists were not willing to accept the infection hypothesis for gastritis or stomach cancers. It was a long-held notion that since the stomach juice is highly acidic, microbes did not survive in the stomach, much less cause infections of the stomach. Marshall and Warren's proposal also lacked a direct proof: Nearly a century earlier, Robert Koch had laid down basic tenets, known as "Koch's postulates," to prove an association between an organism and an infection. One rule required for a clear demonstration of the microbe producing a disease when injected into animals or humans.

So, Marshall needed to prove that the microbe cultured in his lab actually caused gastritis. Had he required permission for his planned study, his Institutional Review Board would have declined. But as Marshall recalled later, he managed to recruit "a 32-year-old light smoker and a social drinker" as his first experimental subject: himself.

First he requested that his colleagues obtain a biopsy from his stomach to make sure that it was normal. After a month of recovery, he fasted overnight and ingested a medicine (cimetidine) that soothed and neutralized his stomach acid.

At 11 AM the next day, as he recalled later, he "swallowed the growth from a flourishing three-day culture of the gastric biopsy-isolated from a 66-year-old patient who had dyspepsia." Marshall had decided not to take any medications even if he fell ill from the ingestion of the culture material containing the gastritis-producing microbes. He began keeping a diary, noting the signs and symptoms as they evolved.

For a whole week he suffered from severe stomach pain, nausea, and diarrhea. On the tenth day, he asked his colleagues to obtain a second biopsy from his stomach. On the 14th day, upon his wife's insistence, he started antibiotics and recovered from the self-induced gastritis.

The culture from the second gastric biopsy of his stomach also revealed classic features of gastritis and a robust growth of the microbes Marshall had swallowed, fulfilling all of Koch's postulates. The microbes were later named *Helicobacter pylori*.

The practical consequences of Warren and Marshall's discoveries were impressive. The most common treatment for severe gastritis that could not be controlled with antacid therapy was the removal of part of or the whole stomach. But, based on Warren and Marshall's research findings, doctors began treating gastritis with common antibiotics for a few weeks, which completely cured gastritis and reduced the occurrence of stomach cancers: soon, surgical removal of the stomach became a thing of the past.

Similar to Beaumont's study of physiology of the stomach acid production 160 years before, Warren and Marshall's discoveries in the 1980s were important in understanding the cause of gastritis and stomach cancers. In a survey, 42 of 50 scientists named the discovery of *H. pylori* and the cure for its associated diseases as the leading digestive research finding "of the millennium."

In 1995, Marshall received the Albert Lasker Award and, in 2005, Warren and Marshall won the Nobel Prize in Physiology or Medicine.

The Life of the Flesh

The belief that the blood is sacred and sanguine appears to date back to prehistoric times. Yet, paradoxically, bloodletting, not blood transfusion, dominated as the first line of treatment for all sorts of ailments. Historians are not sure why and when bloodletting became a part of medical treatment. One theory holds that since people believed that "bad blood" caused illnesses, letting it out made sense.

Because doctors as well as patients believed that bloodletting was the most effective treatment, injecting foreign substances, including blood, was an unthinkable undertaking until the 17th century. In 1657, Christopher Wren (1632-1723) inaugurated the modern era of intravenous treatment when he used quills as needles to "open a vein" in dogs, and administered opium with pumps he had invented. His intent was simply to test his pumps—not to treat the animals with drugs intravenously. Without his knowledge, Wren had invented the first syringe for injecting intravenous medications.

A decade later a French doctor Richard Lower (1631-1691), wondered that if, as the Bible said, *the life of the flesh is blood,* would it not make sense to restore life by infusing blood rather than letting it drain? But he did not know how to do this. First he began experimenting with animal-to-animal transfusions using Wren's technique of opening veins and forcing blood into the recipient animals using pumps.

In 1667, Jean Baptiste Denys of France transfuses blood from a "gentle lamb" to a 16-year-old youth.

Another French doctor, Jean Baptiste Denys (or Denis, 1643-1704), advanced Richard Lower's idea and transfused nine ounces of blood "from a gentle lamb" in 1667 to a 16-year-old youth, who was "tormented with fever." Denys wrote that the boy felt "a great heat," became very ill, and almost died—but did not.

French doctors were unhappy that Denys had experimented on

a sick boy. But Denys continued transfusions using animal blood into humans. He explained that animal blood was "purer than man's' blood" because the latter often indulged in "debauchery and irregularities of eating and drinking."

Some years later, more doctors began transfusing blood from "gentle" lambs and "innocent" calves, sporadically through much of the 18th century. The practice gradually fell out of favor when doctors began to see that many of their recipients died soon after receiving animal blood transfusions.

The London obstetrician James Blundell (1791-1878) gets the credit for attempting the first human-to-human transfusion. He decided that only human blood should be used for human transfusions; in 1818, using instruments, syringes, and funnels he had developed, he began transfusing blood directly from donors into the recipients. According to the statistics he provided, five of his first ten patients receiving direct transfusions survived.

Human-to-human transfusions were almost always acts of desperation. Doctors considered direct transfusions only when all other treatments failed. To their surprise, they were noting that even after human blood transfusions, sometimes—but not always—their patients died from violent reactions. The reasons for such complications remained elusive.

As recently as the early 20th century, a variety of liquids were used as *blood substitutes* and were being injected into the bloodstream of patients. Milk from goats, cows, and even from lactating women were considered blood substitutes to be injected into the bodies of desperately ill patients for a cure.

One of the most unlikely blood substitutes was blood obtained from cadavers.

In the early 1930s, in Moscow, Dr. Serge Yudin, an emergency room physician, had to treat a patient who was bleeding profusely from a traumatic wound. Six hours earlier, an old man had died in the same emergency room, and his cadaver was still in the morgue. Yudin knew that arterial blood did not clot after death. Out of desperation, he withdrew several ounces of blood from the cadaver and transfused it to the bleeding patient, who recovered from shock. Upon learning this discovery, the State Attorney was elated; he permitted Dr. Yudin to collect and preserve blood from cadavers for future transfusions.

Ambulances loaded with cadavers arrived at Yudin's "blood stations," as they called them. Yudin transfused cadaver blood into more than 2,500 patients over the next few years.

Doctors in the United States learned of cadaver blood transfusions, but the idea of transfusing cadaver blood into living patients did not catch their fancy. Bernard Fantus (1874-1940), a surgeon at Chicago's Cook County Hospital, hit upon a new idea. He extended the Soviet idea of cadaver blood transfusion, but considered storing the preserved blood obtained from living humans, or banking the blood.

By 1937, Fantus discovered that a mixture of a pint of blood with 2.5% sodium citrate solution prevented blood from clotting. It could then be stored in refrigerators safely up to ten days. He started collecting blood from healthy donors and storing them, with a clear logic that saved blood could be used in emergencies when a donor was not immediately available.

He thought that this was like depositing *money in the bank* and withdrawing it for later use. So, Fantus named his Blood Preservation Laboratory *The Blood Bank*. On March 15, 1937, a few months before the United States plunged into the Second World War, the world's first blood bank opened at Cook County Hospital's *Fantus Clinic*.

If blood transfusions were difficult in adults, transfusing children, infants, and the newborn was almost impossible. By the early decades of the 1900s, with the introduction of incubators, more newborn infants were surviving, and some of them suffered from severe anemia due to a condition called "hemorrhagic disease of the newborn."

Until later research showed that this condition occurred due to a deficiency of vitamin K, a single injection of which prevented bleeding, transfusion was the only option to save such babies. Transfusing blood into babies was very difficult for other reasons as well: the most basic tools needed for baby transfusions, such as syringes, needles, catheters, tubes, and pumps in appropriate sizes, were not available until the mid 1940s.

In March of 1908, a term infant was brought to a New York hospital. On the second day of birth, she started bleeding in her stools, and then began to bleed under her skin and around her belly button. Small oozing continued for four days, and she appeared to be horribly sick. The infant's doctor, Samuel Lambert, wrote in the medical records:

"As the child's skin became waxen white and mucous membranes without color, it was decided to attempt transfusion of blood obtained from the infant's father."

Lambert summoned surgeon Alexis Carrel from the Rockefeller Institute. Carrel had been experimenting repairing torn blood vessels and connecting them in experimental animals. Upon Lambert's request, Carrel decided to surgically connect the father's artery to the baby's vein. The father's radial artery—the artery at the wrist—was connected to the baby's popliteal vein in the groin using a surgical method called end-to-end anastomosis.

The baby and the father were not given any anesthetic medications during surgery, since none were available. The father's blood group was not matched with his daughter's since few doctors knew about blood groups, let alone how to test for compatibility.

Neither Carrel nor Lambert knew how much blood must be transfused into that little baby's body. They also could not tell how much of her father's blood was flowing through the artery-to-vein connection. Lambert wrote later,

"...enough blood was allowed to flow into the baby to change her color from pale transparent whiteness to brilliant red...her pulse became full, and as soon as the wound was sutured, the infant fed ravenously and immediately went to sleep."

Eight weeks later, she had recovered and was discharged home. In spite of Lambert's dramatic report, direct parent-to-infant transfusion did not catch on. Blood transfusions continued to be hazardous, because of unexpected reactions among the recipients, even when the blood was properly matched for major blood types.

In 1901, a German doctor, Karl Landsteiner had discovered major blood types for which he was awarded the 1912 Nobel Prize in Physiology or Medicine. In 1940, he and his associate Alexander Weiner (1907-1976) discovered a blood-group factor which they called the Rh factor, because it was also found in rhesus monkey blood. Landsteiner and Weiner did not immediately appreciate the clinical significance of the Rh factor.

Philip Levine (1900-1987), another scientist independently sorted out the role of Rh factor in a condition known as *erythroblastosisfetalis*, which caused edematous swelling of the body from severe anemia and heart failure in fetuses and newborn babies. The following cohesive hypothesis emerged to explain erythroblastosis fetalis: if the mother's blood group is Rh negative and her first fetus' blood group is Rh positive (because the father is Rh positive), then the cells from the fetus entering the mother's circulation sensitize her to produce antibodies. In her next pregnancy with an Rh positive fetus, those antibodies enter the fetal circulation, destroying fetal red cells.

This knowledge helped doctors in the 1950s to introduce treatment for newborn infants with erythroblastosis. A few decades later, strategies were developed to prevent the mother from becoming sensitized with Rh positive red cells. Now doctors routinely test the Rh blood group status in pregnant women during their first prenatal visit. Among those with an Rh negative blood type, the infant's blood group is tested soon after birth.

If the infant has an Rh positive blood type, the mother is given a medication called $Rh_0(D)$ immune globulin (e.g., RhoGAM® or a similar product). These are antibodies that detect and destroy any Rh positive cells that might have escaped from her fetus into her circulation during pregnancy or during delivery. Thus, the Rh negative mother will be protected from getting Rh sensitized. This is repeated after each Rh positive child an Rh negative mother delivers, preventing her future Rh positive fetuses from developing erythroblastosis fetalis.

Landsteiner won the 1912 Nobel Prize in Physiology or Medicine for the discovery of major blood groups; but many scientists felt that the discovery of Rh subtype also deserved a Nobel Prize. But Weiner and Levine openly fought to be recognized as the first to have discovered the Rh factor in the blood. Their continued disagreement led the Nobel Foundation to hesitate awarding its prestigious prize to any of the three scientists pivotal in discovering the Rh blood types.

Largely due to preventive efforts, the dreaded erythroblastosis fetalis has almost disappeared. For some infants who are still affected with erythroblastosis, doctors perform exchange transfusions. In this procedure, doctors insert a catheter into the large umbilical vein of the baby and withdraw small amounts of blood and replace the same amount of fresh donor's blood. This is repeated until most of the baby's blood, which contains antibodies against the baby's red blood cells, is replaced with that of the donor's Rh negative blood.

Thus, on rare occasions, doctors reenact the ancient ritual of bleeding—getting rid of the "bad" blood, but with an added bonus of replacing it with "good" blood.

To Hear Well Is to See Well

She was young, plump, and huffed and puffed from a heart ailment. To make a diagnosis of her illness, her doctor took her history, felt her back, and tapped on her chest. The final part of the examination required him to conduct "medical auscultation," the disgusting custom of placing one's ear directly on a patient's bare chest to listen to the heart sound. Her doctor was reluctant to undertake this final diagnostic step because he feared that being of "extreme fatness," there would be little use putting his ear on her chest, for her body fat would muffle heart sounds and heart murmurs. Moreover he was very modest; and he had observed that this patient was literally "lousy" with lice on her body.

Suddenly, he remembered a children's game in which one boy tapped and scratched a hollow tree log while other boys took turns listening at the other end.

A brilliant idea struck the doctor: he found a quire of paper, rolled it into a cylinder, placed it on the patient's chest, and listened. To his delight, he heard her heart sounds most clearly—much better than in all his previous auscultation experience.

René Laënnec invents stethoscope

The first stethoscope had been invented. The episode took place in a Parisian hospital in August 1816, and its inventor, René Théophile Laënnec (1781 - 1826), had given the world of medicine an enduring symbol.

Laënnec was born in Quimper, a small fishing port on the coast of Brittany in France. His family was poor, but they were well known for their simple, honest living. When René was three, his mother died of tuberculosis. His two uncles raised him in the relative tranquility of Nantes, sparing him the horrors of the French Revolution. His uncles also supported René through his medical education, and nurtured his lifelong passion for poetry and music.

René was lucky to have a teacher, Professor Jean Nicolas Corvisart, a celebrated physician in the court of Emperor Napoleon. Fifty years earlier, the Viennese doctor Leopold Auenbrugger had introduced chest percussion, the art of diagnosing lung pathology by taping the chest. Corvisart, a super clinician who made diagnoses based on clinical signs and symptoms, had popularized Auenbrugger's chest percussion method in clinical practice. Corvisart had once remarked that "a ward filled with patients is *the* book of knowledge—far more difficult to read than any printed one."

Laënnec put into practice Corvisart's teachings and tried to diagnose patients' pathology by systematic examination and the direct study of his patients at the bedside.

After he had first developed his new "hearing aid," he tried to perfect it; after testing various designs, he settled on a 18-inch long, two-centimeter diameter wooden tube; a funnel-shaped distal end was threaded onto the stem, which in turn fitted onto the proximal earpiece.

Laënnec used the new instrument regularly and began correlating physical findings with the pathological anatomy. He realized that the instrument was like a new eye, aiding one to "see" inside the patient's chest. He named it "stethoscope"—from the Greek "stethos" for "chest" and "skope" for "I see it." In 1819, he published a classic treatise on auscultation; clinical science at its rational best began to take shape.

Even in the liberal atmosphere of post-revolutionary France, the stethoscope and indirect auscultation remained controversial. Doctors in America and England, however, greeted it with much enthusiasm.

In 1823, John Forbes published an English translation of Laënnec's treatise on stethoscopes and their use for diagnosis. Several engineers began modifying stethoscope design. The final design that stayed on was the one with dual earpieces connected to a long rubber tube ending in a chest piece.

For a number of years, the London doctors used stethoscopes that were very long, which they hid under their hats.

In Arthur Conan Doyle's short story *A Scandal in Bohemia*, detective Sherlock Holmes identifies his visitor as a doctor "by the bulge on the side of his top hat!"

Elementary, of course.

Despite its eventual acceptance, Laënnec did not live long to see the success of his invention. Chronically ill and frail, Laënnec died at 45 from the ravages of tuberculosis.

Pudd'nhead Science and Heredity

> *"Every human being carries with him from his cradle to his grave, certain physical marks... his physiological autograph... [which] consists of the delicate lines or corrugations with which Nature marks the insides of the hands and soles of feet..."*

> —*Pudd'nhead Wilson (Mark Twain, 1894)*

Nurses obtain footprints from newborn infants within a few minutes after birth. Those nostalgic treasures end up in baby-books, which is just as well, because the footprints obtained in the newborn period are certainly not "physiological autographs," as Mark Twain claimed.

A study by the FBI fingerprint experts concluded that a later match using footprints obtained in the neonatal period was possible in only one-half of the tested babies—a fifty-fifty chance. Fingerprints obtained later in life, on the other hand, are integral to the science of identification in the criminal justice system.

Using handprints as an individual's mark of identity probably originated in ancient India, where even today, thumbprints are accepted in lieu of signatures from individuals who are unable to sign their names. The notion that digital marks can also be used to match people began to develop shape in the West only with the advances in population genetics and statistical sciences.

Sir Francis Galton (1822-1911), the brilliant Victorian polymath, logician, and philosopher, was also a social scientist and population geneticist. He developed new ideas to study heredity, and contributed to several basic concepts in statistics. While his contributions to statistics are famous (such as the discovery of "regression to the mean" phenomenon), his contributions to fingerprinting and forensic science are not so well known.

Galton had an enduring interest in the study of the mysteries of heredity. Over a five-year span, he collected handprints from men and women volunteers from various ethnic and racial groups and age categories in England. His sample included members of the same families, from different but closely related families, and from sets of twins and triplets.

After analyses of the collected fingerprints, Galton concluded that even among the closest of relatives, individual variations were large enough to characterize unique patterns of ridges and grooves (he called them *minutiae)* to enable identification. To explain similarities and differences occurring by chance, he proposed a mathematical test, which was taken up by Galton's student and admirer Karl Pearson, who developed the *chi-squared test* eight years later.

Galton's initial interest was to see if patterns of fingerprints were inherited. He discovered that similar patterns occurred more often among relatives than non-relatives. He thus made a general conclusion concerning the individual's specificity of fingerprints as marks of identification. He did not declare that no two sets of fingerprints will be alike—but in a series of papers and books, he presented the *statistical probability* or rarity of an exact match of fingerprints between individuals.

Thus, Galton's work helped place the art of studying fingerprints on a firm scientific basis. It took many decades for other scientists to understand and use Galton's findings, especially into criminal forensic science.

Mark Twain, a Galton contemporary, was also at the cutting

edge of fingerprinting science. In his 1883 biography, *Life on the Mississippi*, Twain wrote:

> *"When I was a youth, I knew an old Frenchman who had been a prison-keeper for thirty years, and he told me that there was one thing about a person which never changed from cradle to the grave—the lines in the ball of the thumb; and he said that those lines were never exactly alike in the thumbs of any two human beings."*

In his devastating masterpiece *Pudd'nhead Wilson*, Twain's protagonist, David Wilson, "a homely, freckled, sandy-haired young fellow," comes to Dawson's Landing, a small town along the Mississippi. He has a "strange" hobby: he likes taking fingerprints from anyone he encounters. Early in the story, he obtains fingerprints from a little baby, supposedly the son of a slave woman; the same fingerprints turn up decades later to solve a murder mystery.

In Dawson's Landing, no one believed Wilson's soft, *pudd'nhead science* of fingerprinting—so they called him *Pudd'nhead Wilson*.

Now few doubt the Pudd'nhead Science.

Dr. Charles Alexander Eastman

Figure 1. **Dr. Charles Alexander Eastman**; Reproduced with permission, Minnesota Historical Society, St. Paul, MN

On March 15, 2021, the United States Senate confirmed Rep. Deb Haaland (D-NM), a member of the Laguna Pueblo Tribal

Nation, as Secretary of the Department of Interior. This historic action marks the beginning of an end to centuries of invisibility of Native Americans in high-profile government positions. Even previously famous Native Americans have faded into collective amnesia.[1]

Consider Dr. Charles Alexander Eastman (1858–1939), a pioneer Native American doctor, known as "Ohiyesa" to his family and to the members of the Santee Sioux Native American tribe (Figure 1 and 2). He graduated from an Ivy League college in the nineteenth century and obtained an MD from a US medical college.[2] A reformer, pioneer, and an author of nearly a dozen books,[3-12] Eastman was the topic of a biography,[2] a documentary,[13] and the focus of the 2007 HBO film *Bury My Heart at Wounded Knee* based in part on a 1970 nonfiction book.[14-15] One of the most famous Native Americans during his time, Eastman is hardly remembered by the medical community today.

Ohiyesa's maternal grandmother, a Medwankton Sioux, married well-known artist Captain Seth Eastman. Their daughter, Mary Nancy Eastman, married Chief Many Lightnings, a Wahpeton Sioux in 1847. Charles was their fifth child. Sadly, his mother died following his birth. At age four, he was given the Native Indian name "Ohiyesa" (the Winner). During the Sioux Uprising of 1862, he got separated from his father and fled from Minnesota to Canada with his grandmother and his father's brother Mysterious Medicine (whose other names were White Foot Print, Big Hunter, and Long Rifle).[14] As Ohiyesa grew up, he learned the ways of Native American culture and heritage. When he was fifteen, he reunited with his father, returning to their home in today's South Dakota.

Ohiyesa, renamed Charles Eastman,[2] earned a BS from Dartmouth College in 1887 and an MD in 1890 from the Boston University School of Medicine. He was the third Native American to earn a medical degree in the United States—after Dr. Susan LaFlesche Picotte of Omaha, who was the first, and Dr. Carlos Montezuma from Chicago, who was the second; both received their

MDs in 1889. Eastman took up employment in the Office of Indian Affairs later that year and was assigned to work as a physician at the Pine Ridge Agency in South Dakota.

Then, there was the Wounded Knee Massacre on December 29, 1890. Although he was not an eyewitness to the massacre, he arrived at that site on New Year's Day 1891 despite a blizzard and the US Army's efforts to delay him. To his horror and sorrow, he saw the scattered remains of the dead and wounded who were (as he wrote later) ". . . relentlessly hunted down and slaughtered while fleeing for their lives." He added, ". . . When we reached the spot where the Indian camp had stood, among the fragments of burned tents and other belongings, we saw frozen bodies lying close together or piled one upon another. It took all my nerve to keep my composure in the face of this spectacle, and of the grief of my Indian companions, nearly every one of whom was crying aloud . . ." He was the only doctor available and did his best to treat the wounded and helped to bury the dead.

Figure 2. **Portrait of Dr. Charles A. Eastman (Ohiyesa), at the estate of Mr. Ward Burton, Lake Minnetonka.** Reproduced with permission, Minnesota Historical Society, St. Paul, MN

Eastman continued working as a reservation physician until 1903. He was the president of the Society of American Indians following World War I. During the 1920s, he served as an inspector of the conditions of Native American reservations. He died on January 8, 1939.

Eastman has been hailed for his remarkable ascent within the ranks of white society's social and administrative domains and for extraordinary contributions to educate and preserve Native American identity. He accomplished these through a set of brilliant writings (assisted by his wife Elaine Goodale Eastman), lectures, and government work. He also cultivated relationships with legislative branch members and served on key committees, such as the Committee of One Hundred Advisory Council, which recommended reforms on federal Indian policies.[14] He encouraged outdoor education and worked as Indian secretary for the YMCA. Later he helped found the Boy Scouts of America. Between government posts, he continued to practice medicine while being active with his lobbying efforts. The cumulative effect of such efforts increased acceptance of Indian culture and rights in the general population.

Dr. Eastman was a pioneer, social activist, and a man with a vision and a noble soul. He tried to educate society about the contributions of Native Americans to American culture and civilization. Yet, he is hardly remembered these days.

This can be rectified, as the selection of Ms. Deb Haaland as the Secretary of Interior has attempted to rectify past wrongs. In her role, she will supervise the operations of one-fifth of the United States land mass which once belonged to Native Americans.

To honor Dr. Eastman, professional medical societies can create awards or scholarships in his name, and his biography could be included in the curriculum of medical colleges, encouraging generations of Americans to learn the legacy and contributions of Native Americans to all Americans.

Acknowledgment: I wish to thank Dr. Gail Johnson, Ms. Diane D. Evans, and Dr. Jeffery Evans for their help with making sure the accuracy of the contents of this paper.

References

1. Reclaiming Native Truth, A Project to Dispel America's Myths and Misconceptions. https://rnt.firstnations.org/. Published 2016-2018. Accessed January 24, 2021.
2. Wilson, R. *Ohiyesa: Charles Eastman, Santee Sioux.* Urbana, University of Illinois Press 1983.

3. Eastman, CA. *Memories of an Indian Boyhood*, autobiography; McClure, Philips, 1902.
4. Eastman CA. *Indian Boyhood*, New York; McClure, Phillips & Co., 1902.
5. Eastman CA. *Red Hunters and Animal People*, legends; Harper and Brothers, 1904.
6. Eastman CA. *The Madness of Bald Eagle*, legend; 1905.
7. Eastman CA. *Old Indian Days*, legends; McClure, 1907.
8. Eastman CA, Goodale-Eastman, E. *Wigwam Evenings: Sioux Folk Tales Retold, legends*; Little, Brown, 1909.
9. Eastman CA, Goodale-Eastman, E. *Smoky Day's Wigwam Evenings*, 1910.
10. Eastman, CA. *The Soul of the Indian: An Interpretation*, Houghton, 1911.
11. Eastman CA. *Indian Child Life*, nonfiction, Little, Brown, 1913.
12. Eastman CA. *Indian Heroes and Great Chieftains*, Little, Brown, 1918.
13. Beane, K. The Vision Maker Media documentary *OHIYESA The Soul of an Indian* (2018).
14. HBO film *Bury My Heart at Wounded Knee* (2007).
15. Brown, D. *Bury My Heart at Wounded Knee*, 1971 New York: Holt, Rinehart & Winston, 1970.

Acknowledgement: This article was previously published in *Hektoen International*, **Volume 13, Issue 3– Summer 2021**. I thank the editorial team for the permission to reproduce it. I also thank those entities that gave copyright permission to reproduce the figures in this article.

The First True Anatomist

In August 1542, a caravan of donkeys trekked along the Italian Alps toward Basel carrying 250 woodcuts and a manuscript securely bound for the arduous journey. Its young author, Andreas Vesalius (1514-1564), journeyed later, and published *De Humani Corporis Fabrica Libri Septem* in Basel, in June 1543, and presented a copy to the Spanish Emperor Charles V (1500-1558).

As reward for his work, the author sought and received the post of "Imperial Physician" at the court. Secure under royal patronage, Vesalius distanced himself from Padua, where his critics were seething because of his anatomy teachings, radically different from those of the great Claudius Galen, a 2nd century Greek doctor, who was never to be contradicted.

But neither Vesalius nor his royal employer expected that the book would become a priceless gift to mankind, shattering the rubbish being taught as anatomy and becoming a landmark in modern science.

When Vesalius was training to become a doctor, his anatomy teachers did not dissect cadavers or make distinctions between human and animal anatomy. They literally followed what Galen had written. Sitting in their balcony chairs, they read from Galen's books while a barber dissected the cadavers of dead animals in the center of the hall, and a "demonstrator" in anatomy pointed out the body parts to mostly bored and sleeping students.

But even while studying anatomy as a student, Vesalius did not like the education he was getting. He wanted to understand human anatomy by studying the human body—not by dissecting other animals. But he had no access to cadavers. So, at night, he began robbing graves of executed criminals, carrying them home and meticulously preparing them for systematic dissection. Working from sunrise to midnight, he examined each organ, studied the connections of each muscle, measured the length of each nerve, noted the features of each bone, and kept meticulous notes on all.

In just four years he was ready. He compiled his discoveries in a volume, titled *De Humani Corporis Fabrica;* "The Fabric of the Human Body." The book talks about the very *fabric* with which the body is constructed and put together.

The *Fabrica* was the first book based on massive research conducted by a single man and completed to near perfection. To complement the text, Vesalius had commissioned a famous Renaissance artist Jan van Kalkar. Such illustrated textbooks of anatomy were very rare. Even the introductory volume *Epitome* contained eleven plates of drawings.

Most anatomists welcomed Vesalius's book. But some attacked it because it questioned Galen. Teachers devoted to Galenic writings, which had also been accepted by the Church, did not wish to change their own beliefs.

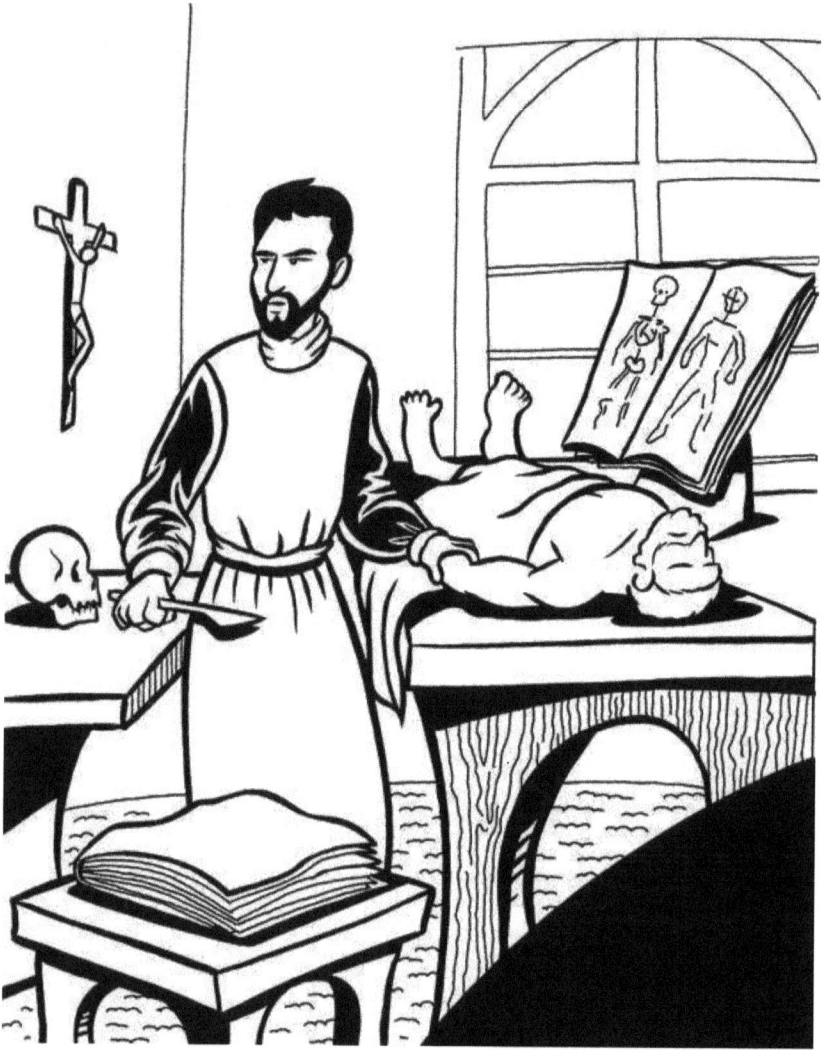

Andreas Vesalius

For example, it was taught that since God created Eve from Adam's rib, men had one less rib than women. But Vesalius said that by counting he had proved that the rib counts were the same between men and women. Outraged traditional anatomists, such as Vesalius's teacher Jacobus Sylvius (1478-1555), pronounced it was the moral failings of contemporary men that accounted for the equal rib counts between men and women. Other anatomists were angry that the illustrations in Vesalius's book "distracted and diluted" students' learning.

To quell such attacks, Vesalius arranged for public dissections and demonstrated over 200 differences between the skeletons of apes and humans. Although some opposition persisted, it did not take long for the acceptance of his anatomical teachings as the standard for centuries to come.

Vesalius died in 1564 en route to the Holy Land, perhaps in a shipwreck. But the reasons for his pilgrimage are not clear. According to one story, a female cadaver Vesalius was dissecting suddenly "came to life," and rumors spread that the famous doctor had murdered the noble woman. In penance for this crime, the great scientist was on a pilgrimage.

On March 18, 1998, Christie's in New York sold a large collection of volumes of The Haskell F. Norman Library of Science: there was one priceless book—the original *Fabrica* Vesalius dedicated to Charles V on or about August 4, 1543.

In one of the illustrations in *Fabrica* there is a drawing of a tomb with an inscription that reads, "Genius lives on, all else is mortal."

Vesalius lives on in his work.

Rays of Hope

On November 8, 1896, a professor of physics at the University of Würzburg was experimenting with electricity and noticed something unusual.

Instead of complete darkness, there was a faint green glow. "Torn between doubt and hope," he recalled later, he repeated the experiment 17 times, and 17 times he produced the same light shimmering through almost anything he placed in front of it.

Now he was convinced that these were "a new kind of rays." Since he did not know their nature, he named them the "unknown rays," or the "x-rays."

Professor Wilhelm Roentgen (1845-1923) could not have known then that his discovery would change the future of medicine.

When he was 18, Roentgen had experienced a similar excitement, but with a different twist. One day in school, he was caught laughing at a caricature of his teacher, drawn by a classmates of his; the teacher insisted on knowing the identity of the artist, but Wilhelm wouldn't reveal it. He was summarily dismissed from school; but he was thrilled because of the victory of his moral courage.

Dismissal also turned out to be a blessing for this German-born young man, growing up in Holland. He sought private tutoring and secured admission to the University of Utrecht as a special student, and later enrolled in an engineering school in Zurich, Switzerland.

Roentgen graduated in 1868, obtained a PhD in physics in 1869, and ascended along the grooves of academia. In 1888, he went to the University of Würzburg, Germany, as a professor, where he eventually became its rector. Roentgen was a popular teacher and a respected researcher, but his career remained quite mundane.

However, there was plenty of excitement in the world of physics in the late 1880s. Edison, Tesla, Crookes, and Hertz were famous for their work on light. Roentgen's colleague, Phillip Lenard, was a pioneer in cathode rays. While he pursued experiments on cathode rays, Roentgen had stumbled on to his immortal discovery.

Roentgen loved working alone. On that fateful November afternoon in his laboratory, he was testing the opacity of black cardboard sheets to block cathode rays emanating from a large Crooke's tube connected to an induction coil. Before beginning the experiment, he switched on electricity to test if everything was right. Soon those "invisible rays" glowed green on a screen painted with barium platinocyanide placed on a table three feet away.

For the next six weeks, Roentgen isolated himself in the laboratory and reconfirmed that he was not crazy—the new rays were real. The rays would pass through wooden blocks, heavy books, tin boxes, and even human flesh. He asked his wife Anna Bertha to keep her hand in front of the rays for 15 minutes, and obtained a picture—the first x-ray of a human organ was that of his wife's hand, once again proving the adage that behind every successful man there is the hand of a woman.

Wilhelm Roentgen Discovers X-rays

Within days of publishing his discovery, Roentgen became an international celebrity. The German emperor had him demonstrate the mystery of the new rays. Doctors realized the utility of x-rays in clinical medicine. Newspapers across the globe hailed the discovery. Puritans worried that x-rays may invade their privacy. But within one year commercial x-ray machines were being used for diagnostic purposes.

Roentgen was the obvious choice for the first Nobel Prize in Physics in 1901.

Although he was an international celebrity, Roentgen remained modest. He never patented x-rays. He donated all of his Nobel Prize money to the university. But World War I had impoverished Germany; suffering from cancer, Roentgen died poor in 1923, four years after his beloved wife Anna had died.

Elizabeth Blackwell—
Not a Dean's Joke

For centuries women worked as midwives, nurses, and even assistants to doctors and surgeons, but they were rarely given formal admission to prestigious medical colleges in Europe and America.

Elizabeth Blackwell (1821-1910) broke this barrier. She received a medical degree from an accredited college in the United States—a milestone in medical education.

Blackwell was born in Bristol, England, where her father Samuel was a nonconformist who fought slavery. Young Elizabeth and her sister joined their father by giving up sugar in protest against slavery. In 1832, the Blackwell family emigrated to the US, settling in Cincinnati. Elizabeth Blackwell decided to become a doctor and in 1845 applied to many schools; all declined, except for the Geneva College of Medicine in upstate New York.

Even at Geneva College, her admission was a fluke.

The dean, Charles Alfred Lee, had presented Blackwell's petition directly to the student body, expecting a denial to admit a woman—even a single "nay" vote would be enough to deny her the seat. Instead, students thought it was joke the dean was playing. They voted unanimously to let the applicant enter.

Stephen Smith, a student, wrote about the first period of class Blackwell attended:

"A hush fell over...as if each member had been stricken by paralysis... A death-like stillness prevailed during the class..."

While her entry became possible because her fellow students thought it was a joke, the medical training was no joke for Blackwell. The people in the town looked upon her as a freak, and snubbed her whenever they had a chance. They jeered when they saw her anywhere, avoided her on the streets, and refused to rent her an apartment. They were unanimous in their opinion: "a lady medical student is either mad or bad."

Women who faced such stereotypical characterization, yet who wished to become doctors often undertook extraordinary measures to achieve their goals.

Take Dr. James Barry: he was a duly certified surgeon and served with distinction in the British army until 1864—for over 40 years of his life. Many people thought that this short man with reddish curly hair was "odd-looking," yet he was famous for his surgical skills. When he died in 1865, a cleaning woman saw his body on the autopsy table and shouted that Dr. Barry was a "perfect female in all her parts."

But the Royal Army declined to comment: they were not only tight-lipped, but also wrote on the death certificate that Dr. Barry's sex was "Male." But, that Dr. Barry, possibly a female, served in the British Army became a sensational news and a topic for continued gossip, plays, and novels for decades since Dr. Barry's death. Even now many believe that Dr. Barry was indeed one Ms. Margaret Ann Bulkley in real life.

While it is possible that Dr. Barry was the first "lady doctor" to receive a medical degree, in reality, Barry's sexual identity remains a mystery to this day. Some experts think that the person was a hermaphrodite, or suffered from a variant of a condition called Kleinfelter's syndrome. However, it is clear Dr. Barry herself thought she was a woman. It is because of that reason she hid the identity of her sex.

It is remarkable that in such a prejudicial atmosphere against women, Blackwell managed to earn a medical degree. She was a superb student, winning the respect and admiration of her classmates and teachers; she graduated at the top of her class in 1849.

Blackwell did not rest on her laurels: She traveled, studied, and worked in Paris, London, and in many US medical institutions and hospitals. In 1856, she helped found the New York Infirmary for Women and Children, and during the Civil War, she trained hundreds of nurses for the Union Army.

She returned to England in 1869 and helped found the London School of Medicine for Women and served there as professor until 1907.

Blackwell's legacy turned out to be much better than Barry's. By 1910, the year of her death, 7,399 women were practicing medicine in the US alone. By the mid 1990s, 150 years after her entry into medical school, more than one-half of each year's graduating medical class students in the United States were women.

A Great Train Robbery?

As the Moose Party candidate for president of the United States, Theodore Roosevelt was campaigning on the evening of October 14, 1912 in Milwaukee.

Then someone shot him.

Although wounded, TR was saved by the metal spectacle case he had kept in his jacket's inner pocket, along with sheets of paper on which he had written his speech. He was bleeding, but the former president insisted on completing the 50-minute speech.

Later his assistants arranged for him to be taken to Chicago, about 60 miles away, where four surgeons would await his arrival at the railway depot at 8 AM the next day. All four surgeons were from the famous University of Illinois Abraham Lincoln School of Medicine and Rush Medical College.

But, for reasons not known to anyone, the famous patient did not reach his intended destination. Instead, TR was admitted at a different Chicago hospital, under the care of a different surgeon. Around 5 AM of October 15th, TR was admitted at Mercy Hospital, under the care of surgeon by the name of Dr. John B. Murphy of Chicago.

How did the former president of the United States end up at the wrong hospital, that too, several hours prior to his intended arrival at a different hospital?

A mystery: According to one version, someone in Milwaukee later convinced Roosevelt to seek treatment elsewhere—not at the University of Illinois, but at Mercy Hospital under the care of Dr. John Murphy, then very famous Chicago surgeon.

A more ignominious version has it that when the flamboyant surgeon Dr. Murphy learned of the gunshot injury to the famous patient (through his connections in Milwaukee), he sensed that the patient might not come under his care. He soon set in motion his other connections within the railways, and had the train either intercepted at an earlier stop, or worse still, had it diverted at the Navy Pier so that TR could be whisked away to Mercy Hospital Medical Center.

Dr. Murphy was a clever, daring, but careful surgeon. After examining TR and obtaining an x-ray, he saw the bullet lodged in his famous patient's chest just near the heart. It had penetrated only the chest wall, not damaging the lungs or the heart. (Even today, visitors to the Mercy Hospital's archives can see the hospital admission log with a line entry for the new patient, and TR's original chest x-ray film).

Dr. Murphy wisely decided not to operate or remove the bullet. He gave the standard wound care for TR, but issued regular, almost daily press bulletins and sent frequent wire messages to the *New York Times*.

In the eight days of Roosevelt's stay at Mercy, Dr. Murphy's fame had spread over the country.

But surgeons in Chicago were not amused or pleased: they rebuked Murphy bitterly for having "stolen" their patient. They complained to the American Medical Association formally, but there is no record of any action AMA took against Dr. Murphy.

President Theodore Roosevelt

Dr. Murphy was a prolific writer. His surgical textbook *Murphy's Surgical Clinics* and its later editions remained the standard textbook for medical students in the US, Canada, Europe, and Asian countries through the 1960s.

It was widely quoted that when Dr. Murphy asked Roosevelt about any fears he might have about his bullet wound, TR reportedly said, "I've hunted long enough, Doctor, to know that you can't kill a Bull Moose with a shot gun."

None but Osler

The life of Sir William Osler (1849-1919) was filled with contrasts. Son of a Canadian clergyman, he became one of the most famous doctors in the world. Although he made few original discoveries and invented no life-saving medicine, generations of historians and clinicians continue to admire him for his wit and wisdom. He never served in any military, yet the US Navy christened a WW II Liberty ship the SS *William Osler*.

Osler was born in Bond Head, a town on the edge of the Canadian wilderness. He grew up a prankster and practical joker. After a year of learning theology he began "adventures with a microscope," as he would say later, which led him to a career in medicine. He studied at Montreal's McGill University and earned an MD in 1872. Osler then went to Europe and refined his clinical skills. Back at McGill in 1874, he began a medical practice and at 26, he was named Professor of Medicine.

Osler's reputation as a great teacher spread quickly at McGill. Because of an endearing demeanor and compassion towards patients, Osler became a much-sought-after doctor. In 1884, he accepted a professorship at the University of Pennsylvania, only to leave for the newly built Johns Hopkins Hospital in 1889 in Baltimore. Over the next four years, Osler wrote his magnum opus, *The Principles and Practice of Medicine*, perhaps the most influential medical textbook of the period. Its 16th edition was published in 1947. Mr. Frederick T. Gates, a member of John T. Rockefeller's philanthropic staff, was so fascinated by the *Practice* that he arranged for a donation of $1,000,000 to the Harvard Medical School in 1902 as a gift "inspired by Osler's book."

While Osler was at Hopkins his fame spread worldwide. He introduced "clinical clerkships" for medical students during which the students needed to undergo hands-on training programs in wards. He founded many societies. He was the originator of "Journal Club" meetings at Hopkins, where students and teachers brought published scientific papers for discussion—a practice that has continued to this day in almost all medical academic programs.

In1905, at the height of his fame, Osler, 55, left Hopkins, accepting the position of Regius Professor of Medicine in Oxford, England, where he spent the rest of his life.

One of Osler's original discoveries was a clear description of a condition called *polycythemiavera*, an inherited disorder in which the red blood cell numbers increase dramatically. He also described the role of platelets in clotting, and explained many of the key features of an autoimmune disorder called systemic lupus erythematosis.

However, it is Osler's diagnostic approach that became legendary. He was the "natural historian" of diseases. His encyclopedic knowledge and clinical skills were built upon a foundation of pathological anatomy evolved from his study of over 1,000 autopsies he had performed. His contemplative nature and ability to combine medical history, philosophy, and humanism with humor, wit, and compassion also endeared him to his students the world over.

Osler was knighted in 1911. With wife Grace, the great-granddaughter of Paul Revere, he led a happy married life. But in 1917, tragedy struck—Osler's only son, Revere Osler, was killed in action in the Great War. "We are taking the only medicine for sorrow—time & hard work," wrote Osler. On December 29, 1919, he died at 70 from influenza and its pulmonary complications.

Osler had written over 50,000 letters and published 1,628 papers and articles. In1925, the famous Johns Hopkins neurosurgeon Harvey

Cushing wrote *The Life of Sir William Osler,* a 1,400-page biography, which won the Pulitzer Prize. Nearly 1,600 articles and books have been written on Osler and his life. Medical societies and clubs in his name have been established in Canada, England, USA, and Japan. Lectureships, medals, and prizes are awarded in Osler's name throughout the world.

What explains the phenomenon of "persisting Osler"? Has he been all but canonized?

Osler was not always "politically correct" on such issues as the care of the elderly, or his opinion about women. At one point, he did not endorse the idea for the new specialty of pediatrics (yet he served as the fourth president of the American Pediatric Society).

Osler, however, remains highly relevant today. He would have admired the new approach to synthesizing evidence and the concept of *evidence-based medicine*—but he would have cautioned that without reflection, data collected would amount to garbage collected; he would have been glad that the age-old scourges afflicting mankind have been conquered, but would have been very sad at "random acts of violence" that continue to plague our societies; he would have welcomed technological innovations, but regretted the vanishing art of "inspection, palpation, percussion, and contemplation"—the pillars of bedside diagnosis he had taught and popularized.

Osler aphorisms, too, remain immortal: he once wrote that medicine is "an art, not a trade, a calling, not a business, [in medicine] your heart is equally used as your head."

In his 1905 farewell address at Hopkins, he noted a few ideals for a physician:

> *"...to do today's work well and not to bother about tomorrow... to act the Golden Rule... towards [your] professional brethren and towards the patient committed to [your] care... [and] ...to cultivate such a measure of equanimity as would enable [you] to bear success with humility, the affection of friends without pride, and to be ready when the day of sorrow and grief came, to meet it with courage befitting a Man."*

Sir William Osler

A Giant and a Dwarf

John Hunter (1728-93), a Scottish doctor and physiologist, was a daring surgeon. His knowledge of animal anatomy and his ability to relate body structures with functions made him one of the greatest English surgeons of all time. Because he approached surgical therapies based on impeccable logic, he was called a "scientific surgeon." But many of his experiments were bold, if not altogether outrageous.

A famous example: During his time, doctors debated what caused syphilis and gonorrhea, two of the well-known venereal diseases. Since both are sexually-transmitted diseases, they often coexisted in the affected patients. More than a century later, it was to be shown that syphilis is caused by a spirochete—parasite called *trepenomapallidum*, and gonorrhea is caused by an infection caused by the bacteria called the *Neisseriagonorrheae* (or simply, the gonococci).

The symptoms of syphilis included fever and pox-like lesions over the genitals and skin. With no treatment, syphilitic patients ended up with life-threatening complications, including bone and joint destructions and brain infections. Gonorrhea, on the other hand, is an acute condition that causes fever, pain, and burning sensation during urination, along with passing of pus in the urine—all these symptoms subside within a couple of weeks.

Thus during Hunter's times, doctors knew the symptoms of these venereal diseases, but were not sure if syphilis and gonorrhea were the same condition with different symptoms, or were two distinct disorders.

Hunter wanted to solve the mystery once and for all. When he saw a patient with fever who also had a sore on his penis, Hunter took a small amount of pus from the patient's penis sore and deposited it onto his own penis. Over the next several weeks, he carefully noted changes in his body. He developed the well-known signs and symptoms of *both* syphilis and gonorrhea, and thus vehemently concluded that these conditions were one and the same.

Hunter had made a gigantic mistake: The patient from whom he had obtained pus for self-experimentation had been suffering from *both* syphilis and gonorrhea, because of which Hunter had contracted both diseases. Hunter had conducted this experiment more than century before the discovery of bacteria or other microbes as causes for human diseases. Secondly, Hunter probably died from the long-term complication of syphilis—rupture of an aneurysm of the aorta.

Hunter was a notorious grave-robber. As an apprentice to his illustrious brother, William Hunter, an obstetrician, John had the task of procuring bodies for dissection for William to study. John sought assistance from jobless scoundrels, robbed graves, and brought cadavers to his brother's dissection room at night. John's nocturnal activities remained an "open secret" among doctors, and even the public suspected it.

Despite such notoriety, and being a flamboyant personality, John was admired because he was a deft and pioneering surgeon. He became rich and prosperous; with wealth, he acquired exotic tastes, including a passion for collecting rare biological specimens. By the 1780s, Hunter's personal museum housed hundreds of unusual objects. Of all his acquisitions, the manner with which he conducted a "giant" follow-up remains one of the most notorious examples of Hunter's determination to collect rare objects.

In April 1782, Hunter read a story in London's *Morning Herald* about a "...surprising Irish giant" being exhibited in a local cane shop. "...For the nobility and gentry at half a crown a person," the ad said, one could see in a large room, a 21-year-old gentleman, one

Mr. Charles Byrne (also known as O'Brien). At eight feet, two inches, the paper said, "Byrne was the world's tallest man."

Byrne was born in Ireland. At the age of 19, he stood at eight feet and was already famous in the regions. The villagers attributed his frame to the place of his conception, presumably on a high haystack, they argued, because another pair of "tall" brothers in a nearby village had been conceived atop a high haystack by *their* similarly acrobatic parents. Charles Byrne loitered for some years in Ireland before ending up in London in 1782.

The newspaper ad worked superbly—the giant show was a hit. Thousands went to see him, spending one-half guinea a visit. John Hunter was one of the earliest visitors. As soon as he saw Byrne, Hunter determined to acquire the giant's skeleton for his museum. Of course, there was a hitch—Byrne was still alive, and probably would not be willing to relinquish his skeleton.

Hunter dispatched one of his employees, Mr. Howison, to meet Byrne and "make an offer" for the latter's body. The giant was dismayed and horrified: naturally he rejected the offer and kicked Hunter's men out. But he also sensed that the notorious Hunter was interested in his body. Not only did Byrne *not* wish to die, he hated the thought of hanging in Hunter's museum for eternity.

Hunter never took no for an answer. His hired helpers closely monitored Byrne's moves and his health status. Knowing this, Byrne tried hiding for a while, but he needed the publicity and the income it brought. So, there were "giant shows." But over the years, his health began deteriorating, and he sensed that the end was near.

To secure a secret burial, he spent all his savings, hired an undertaker beforehand, and employed professional, experienced corpse-watchers to guard his body after death. He had strict instructions that after death, his body was to be placed inside a lead coffin, taken via the Thames, and buried offshore, in deep sea. Pleased with such watertight plans, the poor giant died and ended the miseries of his life on the night of May 30, 1783.

Through his detectives, Hunter knew all of Byrne's plans. When he received the news of the impending death of the giant, he was ready. He used plenty of ale and money—perhaps 500 pounds—to grease the professional corpse-watchers and body-thieves.

On the night of Byrne's death, they hauled his body to Hunter's waiting carriage and took it to his home in London. The next day, in a well-publicized "open funeral," they took an empty lead coffin that was completely sealed via the Thames, and gave it a sea burial.

Hunter did not wish the public to know of his newest collection, for he feared that angry mobs might attack him and his house for the theft of the popular giant's body. However, Hunter spent a busy, sleepless night, moving Byrne's body through a subterranean passageway in his house to a secret chamber. A huge copper cauldron waited there. Hunter filled it with water, lit a fire under it, and placed the dissected parts of the corpse in the boiling water.

By dawn, the flesh had been boiled away, and he dished out all of the bones. They were discolored due to boiling, but were intact and complete. He reassembled the bones, but fearing a public outcry, he kept the skeleton in a secret location. Four years later, in a letter to his colleague he wrote that he had "lately gotten a *tall man*..." but was not going to put it on display right away.

The next summer he did: Five years after Byrne's death, a giant's skeleton inside a glass case went on display at Hunter's museum. To provide a contrast, Hunter placed a picture of a midget next to the giant skeleton.

All of Hunter's collections became part of the Royal College of Surgeons' Museum in London, including the giant skeleton. In 1909, Arthur Keith, the museum's conservator, studied Byrne's skull on Harvey Cushing's suggestion. By measuring the pituitary fossa, a small cup-shaped depression in the base of the skull, scientists concluded that the famous giant suffered from a tumor—an adenoma—of the pituitary gland at the base of the brain.

The hanging skeleton of Charles Byrne remains a grinning testimony to one doctor's perseverance, tenacity, and scientific curiosity, which may well be criminally unethical.

The Baron of Bedlam

Two hundred years ago the discipline of psychiatry did not exist, even though there were plenty of psychiatric patients for whom life was hell. For the common man the living conditions had gradually improved over decades since the Renaissance, but for the mentally ill things had not changed since the Dark Ages. They were chained, beaten, and thrown into jails. They were bled, purged, and tortured in the hopes of evicting demons from their souls. When such efforts failed, they were burned as witches or warlocks.

In 1547, the City of London acquired St. Mary of Bethlehem, built in 1403 for housing the insane. The term "bedlam" is derived from "Bethlehem." These were "madhouses," built to protect the outsider from the insane rather than to treat and cure the latter. In the 18th century, the genteel and the elite often visited the bedlam for sight-seeing—they were the tourist attractions. Although doctors with some interest in psychiatric illnesses occasionally attended the bedlam and mad asylums, no rational treatment existed.

In the aftermath of the French Revolution, certain things began to change. It was a new age of enlightenment in which the sick and downtrodden began to be considered as human beings. Visionaries initiated philanthropic efforts to combat injustices to the masses.

Among them was a shy, pious, and studious intellectual, Philippe Pinel (1747-1826), who revolutionized the care of the mentally ill with humane treatment and simple acts of kindness. He was born in St. André d' Alayrac, a small village in southern France, and studied theology in hopes of becoming a priest. Instead, he switched to medicine and, like his father and grandfather, became a doctor.

Moving to Paris in 1778, Pinel began tutoring in mathematics and translating Latin works into French. During this time, he nurtured liberal ideas toward the human condition, developed close friendships with Benjamin Franklin and Thomas Jefferson, and had plans to emigrate to the United States, but luckily for France he stayed in Paris.

On August 25, 1793, Pinel was put in charge of Bicêtre, a sprawling asylum in Paris built in the 13th century. For a small fee, thousands of Parisians would crowd its lawns on Sundays to see its miserable inmates. Upon taking charge of this horrid dungeon, Pinel appealed to the authorities for reform and, as the first step, released mentally ill patients from chains.

The task was not easy. He made personal plea to the highest authority, the president of the facility, one Dr. Georges Couthon. Firsthand accounts report that Couthon told Pinel, "Citizen, are you crazy yourself to want to unchain such animals?" To which Pinel replied "...these mentally ill are intractable only because they are deprived of fresh air and their liberty."

After warning about Pinel's own safety, Couthon permitted releasing the inmates; this was done under Pinel's supervision. Perhaps this was one of the greatest episodes in the history of modern psychiatry.

The prisoners were released in small batches, and Pinel's experiment in kindness was a success. One of the first inmates released was an English soldier who, seeing the sun after 40 years of confinement in the dark, exclaimed, "Oh, how good it is!"

Philippe Pinel liberates psychiatric patients from shackles.

In 1795, Pinel was asked to take over the Salpêtriére, an asylum for madwomen. Once again, Pinel applied his theory that freedom of the body and kindness with gentility were prerequisites to cure a sick mind. Pinel managed to train assistants, guards, and nurses to think along these lines, continued to unchain the unfortunate women inmates, and soon showed that public safety would not be compromised from such an act.

An expert mathematician, Pinel maintained notes and data and published two monographs: *Nosographiephilosophique* (on the classification of diseases) and *Traitémédico-philosophique sur la manie* (treatise on madness), which became required reading in medical schools. These accomplishments made Pinel popular during his career, but equally important, he managed to keep his head away from the guillotine (see the next story).

When he granted Pinel permission to release the insane at the Bicêtre, the director of the Commune had cautioned, "I fear that you will be the victim of your own obstinacy."

Was the warning necessary?

Hardly. Once when some Parisian hooligans attacked Pinel, a strong, athletically built man came to his rescue. The rescuer was an ex-soldier, a servant of Pinel, but more importantly, one of the first insane inmates Pinel had unchained years earlier at Bicêtre.

The Flame of Life

More often than not, political upheavals tend to victimize the innocent.

Consider Antoine Lavoisier (1743-1794). He discovered oxygen and showed that breathing it helped "burn the flame of life" as the flame of candle. But this brilliant chemist was caught up in the mass hysteria of the French Revolution, arrested and guillotined.

The story of the discovery of the life-giving substance, oxygen, reads like a Greek tragedy. Earlier to Lavoisier, two others had prepared oxygen, but both missed their chance to "discover" it, for they did not realize the importance of their own discovery. The reason for their blindsight was their own conventional, but twisted, logic.

Until the end of the 18th century, scientists had little clue about the nature and composition of various gases in the atmosphere, and the confusing theories about the gases had reached a chaotic pitch.

Georg Ernst Stahl (1660-1734), one of the greatest chemists of his time, suggested that in contrast to test tubes, in our bodies our "sensitive souls" modified our inner chemistry. To explain this, he proposed a logically convoluted theory, which he named the "phlogiston theory." The theory goes something like this: All combustible substances are composed of ash. When they are combined with fire, they release "phlogiston." In closed spaces, the accumulating phlogiston causes the fire to be extinguished.

Another man intimately involved in the story of discovery of oxygen was the English scientist Joseph Priestly (1733-1804). He followed the phlogiston theory to its logical end, and tried to "restore" the flaming power in air once a candle had burned in it. He first took a closed container in which he burned a candle until the candle extinguished. Then he placed a sprig of mint in the same container and left it for ten days, after which, when he lighted the candle again, it started burning. Priestly concluded, "...plants...reverse the effect of breathing and tend to keep the atmosphere sweet."

On August 1, 1774, Priestly heated some quantities of mercuric oxide in a container and collected the released gas, which was oxygen. In this gas, he placed a mouse, and showed that it lived "much better than in the common air." But since he had firmly believed in the theory of phlogiston, he could not interpret the findings in a new logical manner. To top it off, due to political and personal reasons, Priestly fled England to the United States and died there in 1804.

Antoine Laurent Lavoisier faced a greater tragedy. During the French Revolution, Lavoisier, aided by his lovely wife, was busy experimenting on combustion and the process of respiration. After years of hard work, he had produced a new, "fixed air" by using a red calx of mercury. He called it "eminently respirable gas." It produced acids, and so in 1779, he named it *oxygine*, meaning "acid plus beget."

Lavoisier held important positions and had influential friends; but unknown to him, he also had enemies in prominent places. One Jean-Paul Marat, a physician turned libertarian, brought a list of false charges against Lavoisier, most prominently scientific fraud, tax evasion, and polluting the environment with his experiments. The tribunal found Lavoisier guilty.

Antoine Laurent Lavoisier discovers oxygen.

Before sentencing Lavoisier, it seems that the judge said that the Republic had no need for scientists, to which Lavoisier replied: "This saves me the inconveniences of old age."

On May 8, 1794, Lavoisier and 77 others, including his father-in-law, were guillotined. Lavoisier's body was never found in the frenzy that followed.

Years later, the great French mathematician Joseph-Louis Lagrange (1736-1813) is believed to have remarked: "It took but a moment to cut off that head, though a hundred years perhaps will be required to produce another like it."

Lesser Circulation and
a Nobel Heretic

Many a martyr's blood stained the 16th century, but none more poignantly than that of Michael Servetus, a Spanish physician, theologian, and mystic. Seventy-five years before Harvey discovered circulation, and 150 years before Lavoisier's discovery of oxygen, Servetus discovered pulmonary circulation—how blood from the heart entered the lung and returned to the heart.

In the second century of the CE, Galen, a Greek physician/surgeon practicing in Rome, had proposed that a mixture of blood and air flowed through the arteries, propelled by the heart's pumping. His teaching led to the proposition that all circulation began and ended in the liver, and that the lungs simply fanned the heart to prevent it from becoming too hot. The blood on the venous side of the circulation entered the right side of the heart, and seeped into the left side via pores "invisible" to our eyes. This was the concept of circulation prevalent in the early 16th century as Servetus began studying medicine.

Michael (or Miguel) Servetus Villanovanus was born in 1511. He studied Latin, Greek, Hebrew, theology, and the law. At 19, he had become disillusioned with the immorality of the papacy and the church, and began arguing his theological notions with other biblical scholars. His motives were honest, but his pride bordered on foolish impertinence. Servetus studied medicine in Paris, and then spent 14 years in the French town of Vienne as a successful doctor, but in relative anonymity. He wrote a book titled *De Trinitatis Erroribus* (Errors in the Trinity), which questioned the Trinity; naturally it irritated both the Catholic and Lutheran church leaders.

In 1553, he published his last book, *ChristianismiRestitutio* (Restitution of Christianity). Expecting serious opposition, he published the book under a thinly veiled pseudonym, "MV." But the tactic did not work. He was arrested; but, with luck, good connections, and bribery, he managed to escape the Paris prison the next day. His prosecutors burned his effigy along with all the copies of his book they could find.

However, few realized then that *Christianismi* would find a permanent place in medical history, not because of what it said about the human soul, but because of what it said about physiology. Servetus had done animal dissections and studied the heart and circulation. Based on his animal studies, Servetus had concluded:

> "...the blue blood from [the] right ventricle enters the lungs to 'mingle' with the inspired air, whereupon it becomes of 'fiery potency.' Blood then gets back to the left ventricle via pulmonary veins, not by seeping through 'invisible pores' in the interventricular septum..."

It is unclear why Servetus wrote a one-page-long description of the physiology of heart and lung in a book that was largely an argument discussing Christian doctrines. He might have reasoned that his description of pulmonary circulation was proving that mighty Galen was wrong, despite the acceptance of his writings as final authority by the church. Perhaps this was Servetus' way to show that the authorities were wrong in matters theological, too.

After a few months of his escape, inexplicably Servetus reappeared in Geneva, Switzerland. Perhaps he was expecting that the atmosphere in Geneva would be more liberal, and he and his theological doctrines would receive a friendlier reception than in oppressive France.

Sadly, he was mistaken. On the day of his arrival in Geneva, Servetus was recognized and arrested. John Calvin himself conducted the well-publicized inquisition for two months.

On October 26, 1553, the tribunal pronounced Servetus guilty and denied Calvin's last-minute request for a kinder method of execution—beheading.

The next day, after a procession through the town's narrow streets to the banks of the picturesque Lake Geneva, Servetus was bound to the stakes and burned, as thousands of onlookers stared. All copies of *Christianismi* that they could find were burned along with him. It is said that as the flames engulfed his tortured body, Servetus shouted, "Jesus, Son of eternal God, have mercy on my soul."

Servetus was burned because of his "heretic" religious views, not for his medical discovery. He was a prisoner of his time. It was his naiveté and arrogance that brought him the unwanted martyrdom.

The authorities believed that they had burned all copies of *Christianismi*; but three have survived. One of these is in a Viennese museum, the other in a Parisian museum, and the third, partially burned copy, is in a museum in Edinburg, England. The Paris copy is of particular interest. An inscription by one of Servetus' prosecutors on it suggests that possibly this was the very copy used during his trial. There is also a stain—a smoke and fire mark, possibly suggesting that as Servetus burned, a souvenir hunter plucked it from the burning stakes.

Michael Servetus

He Sewed up a Beating Heart

July 9, 1893: It was a hot evening in Chicago. A simple argument in a South Side tavern led to a brawl, which intensified into a fight. Tempers flared, and a 24-year-old James Cornish was stabbed in the chest.

They took him to the nearby Provident Hospital. Dr. Daniel Hale Williams (1856-1931), the chief surgeon, examined the young man and thought that the wound was superficial. He decided to observe.

But Cornish had a rough night. His chest pain and bleeding worsened, and by morning he was in shock. X-rays had not been invented and good anesthetics and antibiotics were decades away. Even blood transfusions were rare.

Under such conditions, and against the traditional advice of experts to avoid chest surgery for the fear of collapsing the lungs, which would invariably kill the patient, Dr. Dan, as he was fondly called, performed a historic operation upon Cornish.

In the operating room that morning, he extended Cornish's stab wound and cut open two ribs near the sternum. There he made a small window. After retracting the left lung, he saw and sutured about an inch-long tear in the pericardium, the sac around the heart. He saw a small wound in the heart muscle, which he left untouched, hoping it would heal without sutures. He sutured other bleeding arteries nearby, and closed the incisions.

This was a daring operation. But hardly anyone who knew Dr. Dan was surprised that he undertook such a procedure. The good doctor always tried to do the impossible.

After a stormy course, including the collection of "eighty ounces of bloody serum" in his chest cavity, Cornish recovered and went home on August 30th.

Adversity had been Dr. Dan's partner for much of his early life. Daniel Hale Williams was born in Holidaysburg, PA. His parents were black with admixtures of Shawnee and Delaware Indian and many Caucasian strains. They were relatively affluent, but when Dan was only seven, his father died, and his mother moved to Janesville, Wisconsin.

With the help of a benefactor, Mr. Harry Anderson, young Daniel went to high school. He worked part time in Mrs. Anderson's Tonsorial Parlors and Bathing Rooms and sang as a tenor. After high school, he apprenticed under the leading local doctor, Henry Palmer, who helped Dan study medicine at the Chicago Medical College, which was then affiliated with Northwestern University.

Dan graduated in 1883 and set up practice at the corner of 31st Street and Michigan Avenue. He was one of only three black doctors in Chicago.

In 1890, Emma Reynolds, a young black woman, was denied admission in all of the nursing schools in the city. She and her brother, the Reverend Louis H. Reynolds, sought the advice of Dr. Dan. Dr. Dan knew he had no power to influence the medical institutions. So, he and the reverend decided to build a hospital and train nurses and students of all races.

They campaigned and held rallies for their cause. With Dr. Dan's reputation and charisma, support came in from both black and white citizens of Chicago. The Armour Meat Packing Company committed a down payment for a three-story brick house at 29th and Dearborn.

On May 4, 1891, *Provident Hospital and Training School* was founded with an annual budget of $5,429. It was the first interracial hospital in the US. The next year, seven women, including Emma Reynolds, enrolled in its first nursing class. In addition to working at

Provident Hospital, Dr. Dan also held appointments at the South Side Dispensary (1884-92) and the Protestant Orphan Asylum (1884-93).

Dr. Dan's surgery on Cornish, however, remains an American classic. He was not the first to operate on a beating heart. H. C. Dalton of St. Louis had repaired a traumatic heart injury in 1891 and had reported the achievement in 1895. A few others had attempted similar operations with varying degrees of success. But Dr. Dan did not know any of those reports.

Dr. Dan invented many innovations in surgical procedures. One such procedure was the repair of a torn, bleeding spleen. The treatment for patients with a torn, bleeding spleen (usually from trauma) was splenectomy—the complete removal of the spleen. Without a spleen, such patients were at lifelong risk of dying from fatal infections. Dr. Dan successfully sutured a bleeding, traumatically injured spleen without removing it. This was only the second such surgery successfully carried out in the United States.

Dr. Dan received many honors in his lifetime. In 1894, President Grover Cleveland invited Dr. Dan to become the surgeon-in-chief at Freedmen's Hospital in Washington DC, a post he held for four years. In 1913, the American College of Surgeons made him a Fellow and a charter member—the first black surgeon to be so honored. A Chicago Elementary school bears his name. In 1975, the Daniel William Hale house in Chicago was designated a National Historic Landmark.

The Provident Hospital went through financial turmoil and was forced to close in 1987. It reopened in 1993 as part of Cook County's Bureau of Health Services. It later became affiliated with Chicago's Cook County Hospital; this affiliation has continued.

The space on the incorporation papers of the Provident Hospital where Dr. Dan should have signed remains a blank. Just before he was to sign them, Dr. Dan was called away for an emergency—he never got around to completing the signing ritual.

Even though Dr. Dan did not sign the Provident Hospital incorporation papers, his name is etched in the annals of the history of American surgery.

The Legacy of Dynamite

Alfred Bernhard Nobel, the founder of the Nobel Prizes, led a life of paradoxes. His great inventions, blasting caps and dynamite, were almost synonymous with destructive power. Yet, his intent in creating them was to help engineering projects by converting the volatile and dangerous nitroglycerine for safer use. The man who made millions by manufacturing destructive explosives and weapons of war used his wealth to endow everlasting prizes, one of which honors persons or organizations that enhance "fraternity among nations."

The other Nobel Prizes, in literature, physics, chemistry, and physiology or medicine, are testimony to Alfred Nobel's diverse interests. He was an inventor; so, chemistry and physics were part of his life. A poet in his youth, Nobel maintained a lifelong interest in literature. His interest in medicine grew from his scientific mind, but may also have been influenced by his chronic ill health.

Nobel was born in Stockholm in 1833. His father, Immanuel, was an inventor and engineer who had become familiar with gunpowder blasting in his construction projects. Business losses forced Immanuel to move to St. Petersburg, Russia, in 1837, where he succeeded in developing supplies for the army, including naval mines for the Russian military. The family joined him in 1842. Because of his health, Nobel managed only a year of formal schooling in Sweden, but in Russia he received private instructions. By the age of 17 he was fluent in Swedish, Russian, French, English, and German. He had become an accomplished chemist. Then he spent many years traveling, working in chemical labs. In Paris he met and became a friend of Ascanio Sobrero, the inventor of nitroglycerine.

After the Crimean war, Nobel and his parents returned to Sweden. Nobel began to work with nitroglycerine. He invented a practical detonator, which put him on a path that would make him very rich.

His work with the explosives also brought him much financial loss and personal tragedies. In 1864, his factory producing nitroglycerine exploded, killing his brother Emil. Concentrating his efforts on rendering nitroglycerine safe, in 1867, he stuck upon the formula combining nitroglycerine with diatomaceous earth, which made the compound very stable. He named the invention "dynamite."

The invention of dynamite inaugurated Nobel's family fortune: By the late 1870s, Nobel was head of an industrial empire with nearly a hundred factories in Europe and the US. An avid inventor, Nobel held 355 patents.

But this brilliant man spent his last years in isolation in San Remo, Italy, beset by migraines and angina. In 1895, with no legal assistance, he rewrote his will. In it he asked that his holdings be liquidated and the interest from the proceeds used to fund prizes for those who "conferred the greatest benefit to mankind" in the previous year through their work.

On December 10, 1896, Alfred Nobel died alone in his villa.

France, Italy, and Sweden each claimed jurisdiction over his estate—and a share in his wealth. A peculiar French rule settled Nobel's nationality in favor of Sweden. Under the law, a man's "permanent home" was where his horses were kept. Shortly before his death, Nobel had moved his carriage horses to his summer home in Sweden. His family also contested the will, but was eventually won over, with the help of a financial settlement. It took four years of dedicated efforts of two executors to implement Nobel's bequest.

The first Nobel Prizes were awarded on December 10, 1901,

the fifth anniversary of Nobel's death. Except during two World Wars, the prizes have been given out almost continuously since their inception.

Alfred Nobel

A Student for All Times

In 1899, William Sealy Gosset (1876-1937), a 23-year-old chemistry graduate, was offered a job as brewer by the Arthur Guinness, Son & Co. Ltd., in Dublin.

Gosset was born in Canterbury. In school, he proved to be a brilliant student. Because of poor eyesight, he decided not to study engineering like his father, but to study chemistry and mathematics at New College in Oxford, graduating in 1899 with a First Class in chemistry.

His job at Guinness was to apply his chemistry knowledge and use scientific methods and help them produce perfect beer. To brew the perfect beer, one had to mix exact amounts of yeast to the continuously fermenting barley—too little yeast led to incomplete fermentation, and too much to a bitter taste. Unpredictable changes in the ambient temperature also affected the yeast, adding to the complexity of the process.

Gosset's first task was to count the yeast colonies, for which he learned to use the newly discovered hemocytometer, a special glass slide which helped count red blood cells under the microscope. But he did not know how to estimate the quantity of colonies in entire jars based on small samples taken from them. Gosset conceptualized a mathematical and statistical approach, and solved it by developing a simple statistical test.

The scientific basis for Gosset's solution had evolved over 150 years. Mathematicians knew that observations were prone to errors. With improved measuring instruments, errors due to imprecise measurements became smaller; it was then scientists discovered another kind of "error," which they called "random error," especially encountered in biological measurements when identical samples obtained from a large population varied slightly among them. In 1820, Pierre-Simon Laplace (1749 –1827) proposed that random errors (computed by subtracting deviations of the observed from the predicted) can be plotted. Such plots became known as the "normal" or "Gaussian distribution" curves, named after the famous German statistician Carl Fredrick Gauss (1777-1855).

The British mathematician and philosopher Karl Pearson (1857-1936) took the concept of distributions a step further. He proposed that all experiments provided only pieces of information regarding a larger, immeasurable, original population and its scatter. He showed that measurements themselves, not just the random errors, had probability distribution properties, which could be described using four *parameters*, namely the mean, standard deviation, symmetry, and kurtosis.

Pearson proposed the term "parameter," which had a Greek root meaning "almost measurements." Pearson held that if one knew the four parameters, one could locate the probability that an observed number would be at a certain location within the population scatter. He proposed a family of "skewed distribution curves" to describe such scatters. Pearson was a towering personality. He had founded *Biometrika*, a major statistical journal, the first issue of which appeared in October 1901.

Gosset discovered that his yeast colony counts did not fit any of Pearson's "skewed" curves, but fitted the Poisson model, named for Siméon Denis Poisson (1781–1840), the 18th century French mathematician. In November 1904, Gosset presented a report to the Guinness Board titled "The application of the 'law of error' to the work of the brewery," with an intention to publish his scientific findings. Since Guinness Company did not allow its scientists to publish papers, Gosset negotiated an arrangement that allowed him to publish his papers provided he used a pseudonym and did not divulge any confidential data belonging to the company. Gosset agreed.

By 1905, he had met Pearson and the two became close friends. Pearson agreed to the conditions of the Guinness Company and published Gosset's first paper, *On the error of counting with a haemacytometer* under the pseudonym "Student," in *Biometrika* in February 1907. The paper explained how the scatters of colony counts were similar to the exponential limits of binominal distribution.

Pearson had strongly argued that only with large samples could one estimate population parameters. Since most researchers cannot obtain large samples, Gosset thought that formal methods ought to be developed using small samples for estimating population means. He carried out a number of empirical experiments to develop such methods.

In one experiment, Gosset prepared 3,000 pieces of cardboard, on each of which he wrote two sets of data on 3,000 "criminals." One set of values were on heights, and the other, the lengths of the left middle fingers. Gosset shuffled the cards and drew at random 750 samples of four cards each, and computed means and standard deviations of each. Then he obtained the difference between each sample mean and the population mean (n=3000), and divided the difference by the sample standard deviation to obtain 750 values called in statistics the "z scores." He plotted the scores as probability functions and discovered that even without any of the four parameters of Pearson, one could estimate the population mean and the associated error with a fair degree of certainty.

Gosset published these results in his second paper under the title *The Probable Error of a Mean* in *Biometrika* in March 1908, again using the pseudonym Student. For many years this paper remained in obscurity, until Roand A. Fisher, another famous statistician, provided a mathematical proof for Gosset's test and showed how it could be used when researchers were able to study only small samples.

Despite its long algebraic discourses and mathematical arguments, Gosset's second paper is a classic. In lucid English, free from jargon,

Gosset explains the logic of the test. Initially, Gosset called his test a z test to denote the key ratio. Because the convention was to use z for population parameters and t for samples, at Fisher's suggestion, Gosset published another set of tables in 1925 for testing the significance of observations from small samples. This table became the *Table of Student's t Distributions* and the test, *Student's t test*.

For 30 years, Gosset wrote a number of papers on statistics to solve practical problems encountered in the brewery. Yet, who the famous "Student" was remained unknown until his death on October 16, 1937. Some of his friends then solicited and obtained a gift from the Guinness Company to publish Gosset's collected papers, which was done in 1942, revealing the identity of a great student who had taught the rest of world.

Battling poverty, injustice, ignorance and fear, and despair

Figure 1: *Don Quixote de la Mancha and Sancho Panza* (1982), by Maurice D. Pearlman, MD (1915-1985), University of Illinois, Class of 1938. Donated in his memory by his daughter, Martha Pearlman. Assemblage approximately 7' X 11'. This picture was

taken when the statue was on the third floor of the Library of the Health Sciences, University of Illinois at Chicago. A long rope is placed on Don Quixote's horse. Four crutches serve to depict the four blades of the "windmill." On their arms, the inscriptions read: "Poverty," "Injustice," "Ignorance and Fear," and "Despair." A long rope connecting Don Quixote and his horse, lances, and additional small figures surrounding the statue have been stolen.

At the entrance hall of the Library of the Health Sciences of the University of Illinois Medical Campus in Chicago, one can see an ensemble of surgical and anesthetic equipment such as knives, forceps, speculum, towel clips, hemostats, kidney trays, IV poles, crutches, x-ray films, anesthetic balloon bags, and more. They are not discarded materials piled up for recycling—instead, they form an assemblage of meticulously conceived, brilliant artwork. Two transparent, bony figures depict famous fictional characters: a fully armored knight-errant on his horse and his squire riding a donkey. Both are carrying lances and are ready to battle "poverty," "injustice," "ignorance and fear," and "despair," written across the four blades of the "windmill" made of crutches. They are the heroes of the great novel *Don Quixote de la Mancha*, written by the Spanish writer Miguel de Cervantes (1547-1616).

We see Don Quixote riding his horse Rocinante and Sancho Panza, riding his donkey, Dapple[1] (Figures 1 and 2). The top blue banner proclaims "ENLIGHTENMENT" and the middle banner, made of metal that looks white when reflecting light, has an inscription "PEACE" written in twenty-three languages. This banner on the right side (Figure 3) has an inscription that reads "LOVE" in twenty-three languages. The bottom blue banner reads "ONLY HE WHO INTENDS THE ABSURD MAY ATTAIN THE IMPOSSIBLE."

This stunning and magnificent statue created by Dr. Maurice Pearlman (1915-1985), previously located on the third floor of the library,[2] has now been moved to the entrance hall and has been partially restored. It adorns a dark corner of the hallway, hardly drawing the attention of medical students who pass by it.

Dr. Pearlman graduated MD from the University of Illinois Abraham Lincoln School of Medicine in 1939 and served as an ophthalmologist for nearly twenty years before moving to Las Vegas, Nevada. He was not trained in the arts, but admired Cervantes' great novel and its heroes. He was influenced by the Chicago artist John Kearney and managed to create this artwork using a large collection of discarded surgical instruments he had collected over the years of his practice.[2] Pearlman was deeply committed to the humane in the practice of medicine, which is richly displayed in his brilliant creation.

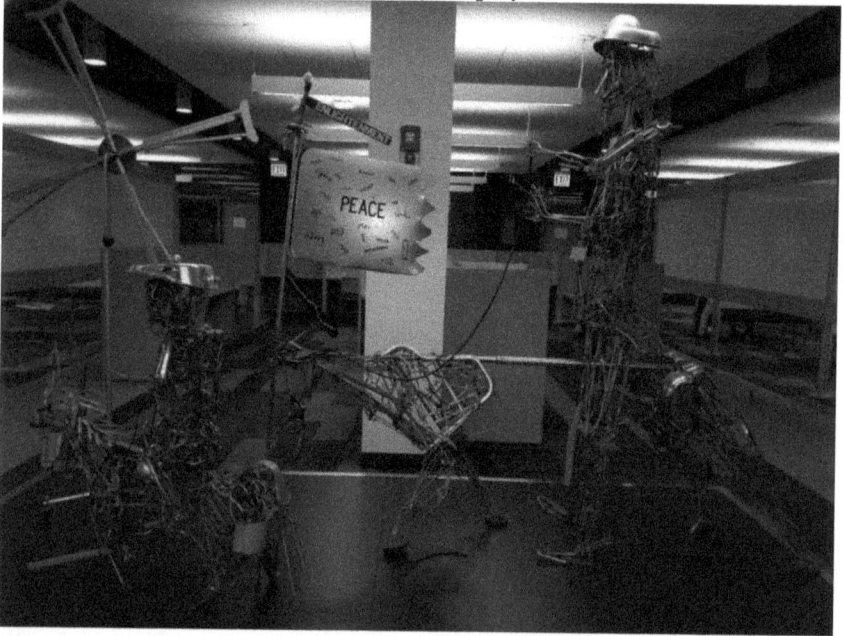

Figure 2: Side view from the left of the statue, now located on the library's main floor. A pole with the while banner has an inscription "PEACE" written in 23 languages. The 'banner' is actually made of metal, which shines white when light is shown on it. The top blue banner has an inscription "ENLIGHTENMENT" and the bottom blue banner has an inscription, "Only He Who Intends the Absurd may Attain the Impossible." The Don Quixote figure is holding the rope attached to his horse.

Figure 3: Side view from the right of the statue. The banner on its other side has an inscription that reads "LOVE" in 23 languages.

Deciphering the allegorical meanings of Cervantes' novel would be beyond the scope of this brief article, but its greatness has been compared to the works of Shakespeare and Dante. In the introduction to the excellent translation by Edith Grossman,[1] Harold Bloom says that it is difficult to appreciate the aesthetic truth of *Don Quixote* and to fully understand the truth of Quixote's quest because ". . . we confront a reflecting mirror that awes us even while we yield to delight." He adds, "Fielding and Sterne, Goethe and Thomas Mann, Flaubert and Stendhal, Melville and Mark Twain, Dostoevsky: these are Cervantes's admirers and pupils."

But what did Pearlman say about the only artistic creation he produced? While arranging his work for donation to the University of Illinois Library of Health Sciences, he noted, "Behold! . . . Don Quixote de la Mancha, defender of our profession against all assailants or detractors, and knight-errant against life's cruelties. This sculpture is dedicated to His (and Medicine's) noble ideals."[2]

Nothing more needs to be added.

References

1. Cervantes M. *Don Quixote: A New Translation* by Edith Grossman. New York: HarperCollins Books; 2003.
2. Scheinman M. *A Guide to Art at the University of Illinois: Urbana-Champaign, Robert Allerton Park and Chicago.* Urbana-Champaign, Illinois.: University of Illinois Press; 1995.

Acknowledgment: I thank the Trustees of the University of Illinois for permission to obtain photographs of the stature shown in this picture, and to Mr. Harold Jaffee, Library Assistant for sending Figures 2 and 3.

Image Information

The sculpture: *Don Quixote de la Mancha* sculpture located on the main floor of the Library of Health Sciences of the University of Illinois in Chicago, Illinois. A plaque at the bottom of the sculpture reads, "This Sculpture by Maurice D. Pearlman, M.D., 1915-1985, University of Illinois College of Medicine, Class of 1938, was donated in his memory by his daughter, Martha Pearlman." The donation occurred in 1995 to the Trustees of the University of Illinois, who granted a permission to reproduce pictures of this statue for the purposes of this paper in *Hektoen International*.

Figure 1: The photograph of *Don Quixote and de la Mancha* statue when it was located on the third floor of the Library of the Health Sciences, University of Illinois in Chicago. Copyright Tonse N.K. Raju, published with permission.

Figure 2: A photograph (side view from the left) of the statute *Don Quixote de la Mancha* as it appears now on the main floor of the Library of Health Sciences, University of Illinois in Chicago. Photograph by Mr. Harold Jaffe, Library Assistant, Library of the Health Sciences, University of Illinois at Chicago, IL, published with permission.

Figure 3: A photograph (side view from the right) of the statute *Don Quixote de la Mancha* as it appears now on the main floor of the Library of Health Sciences, University of Illinois in Chicago. Photograph by Mr. Harold Jaffe, Library Assistant, Library of the Health Sciences, University of Illinois at Chicago, IL, published with permission.

Acknowledgement: This article was previously published in *Hektoen International*, **Volume 13, Issue 3– Summer 2021** I thank the editorial team for the permission to reproduce it. I also thank those entities that gave copyright permission to reproduce the figures in this article.

The Battle of the Snakes

Caduceus, the most widely recognized symbol of medical profession, has two snakes entwined on a rod having a pair of wings. If historians had their say, this symbol would have long been replaced with that of Asclepius (spelt variously, including "Aesculapious"), the Greek god of medicine who carries a long shaft with a single snake entwined on it. Asclepion principles were lofty and noble, but not that of caduceus. It is the "herald's wand," the shaft of Hermes, a Greek god who, among other things, was a patron of thieves and merchants—not of doctors.

Asclepius might well be a legendary name, but it is associated with a physician with great compassion for the sick and exceptional skills as a doctor. Son of Apollo and the nymph Coronis, he was tutored by Chiron the Centaur in the art of surgery and medicine. His abilities as a healer were so great that his fame spread far and wide.

Asclepius is said to have saved many who were on the brink of death, which infuriated the gods of the underworld, who feared depopulation of their lands. They urged Zeus to stop Asclepius from continuing to save the sick. As was the case with many Greek gods, impatient Zeus dispatched a thunderbolt forthwith and ended the life and career of Asclepius prematurely.

Asclepius' fame did not die even after his death. For centuries, Asclepius' cults, including magical cures, were offered in his name.

Hundreds of temples were built with him as the principal deity. Priests and physicians there treated patients coming from all over Europe. Medications, sedatives, and ointments were dispensed, and special diets, exercises, and massages were administered. As they slept in those temples, the patients dreamt of Asclepius appearing with his wand and the sacred serpent, effecting cures and advising remedies. It was believed that Hippocrates was the 18th descendent of Asclepius.

As in other ancient cultures, Greeks looked upon the serpent as a symbol of knowledge and wisdom, healing and good health, and fertility and revival. For 2,500 years, the combined motif of the snake and Asclepius remained synonymous with the noble principles of the craft and calling of our profession.

But, what of caduceus?

The caduceus as the magic wand of Hermes is seen in many Greek artworks dating back to the 6th century BC. Given to him by Apollo, the shaft was made of olive wood, with garlands and a pair of wings.

It seems that Hermes once brokered a truce between two battling serpents by sticking his shaft between them. The snakes ended their fight, entwined peacefully on his shaft, and became the caduceus symbol used today. Hermes is thus regarded as an ambassador, a mediator, and a noncombatant peacemaker who could broker a deal even among battling snakes.

In the 16th century, Sir William Butts, a physician to Henry the VIII, was permitted to use the caduceus symbol on his professional garb. Later, it was adopted as part of a ceremony in the College of Physicians of London, which initiated centuries of confusion about the medical connection of caduceus.

In 1856, the US Marine Hospital Service adopted caduceus to denote the noncombatant status of its medical units. The US Public Health Service used it next, and over the century, hundreds of other professional organizations began adapting the caduceus symbol in their mottos, banners, and trademarks. Many medical journals, societies, and websites use the caduceus symbol. An active list-serve on medical history is called Caduceus-L.

There may be another tenuous, if coincidental, connection between Hermes and today's medical profession. Romans regarded Mercury (Hermes of the Greeks) as a god of sleep and relaxation. Dedicating him the number four, they named the fourth day of the week, Wednesday, in Mercury's honor. It is a charming coincidence that many doctors take Wednesdays off from work for relaxation.

Despite the continued use of caduceus in modern times as a symbol of the medical profession, it is noteworthy that ancient scholars generally held Hermes with considerable disrespect.

In Homer's "Hymn to Hermes," Apollo says of Hermes:

> *This among the Gods shall be your gift…*
> *To be considered as the Lord of those who swindle,*
> *Housebreak, sheep-steal and shoplift.*
> *A schemer subtle beyond all belief.*

Being the symbol of Hermes—the patron of the marketplace, swindlers, and fat cats—caduceus may be an appropriate insignia in today's commercialized medicine. But purists and idealists disagree.

While its continued use by the military-medicine seems appropriate to signify the latter's noncombatant status, caduceus may be the wrong symbol for the medical profession. Historians have suggested substituting it for the more appropriate symbol consisting of a single serpent entwining the shaft of Asclepius. The suggested name for this symbol is the "Aesculapion," or its Greek version *Asclepion*.

The Shaft of Asclepius   Caduceus

The Man Behind the Oath

Medical students take the *Hippocratic Oath* on the day of their graduation. But who was Hippocrates and why is he revered?

Hippocrates, regarded as the "father of modern medicine," was a Greek doctor who lived more than 2,000 years ago on an island called Cos. Very little else of his personal life is known with certainty, but quite a bit is known about what he did as a doctor.

Hippocrates changed how to think about diseases and develop cures. For centuries prior to Hippocrates, the "standard of care" to treat illnesses was to either pray to an angry deity for mercy, or offer attractive gifts to an angry, wicked demon. A sick man was more likely to end up in a temple under the care of a priest than in a hospital under the care of a doctor. If the priest could not cure, a sorcerer or a magician showed up, but not a doctor. Thus there was no distinction between doctors and sorcerers and hospitals and temples.

With his exceptional power of observation and common sense, Hippocrates studied human diseases and insisted that they were from physical causes. He taught that diseases occurred due to derangement within the body, or from forces of nature such as bad weather or an unhealthy diet, *not* from forces in the heavens or the supernatural spirits.

Hippocrates was one of the first to separate medicine from magic. The treatments he proposed were based on common sense and logic.

Using simple household remedies and natural preparations, he cured thousands of people. He abandoned superstitious rituals and magic. His teachings profoundly impacted all of future medical practice.

Hippocrates was a practical man: He did not undermine the power of prayer. "Prayer is indeed good," he said once, "but while calling on gods you must lend a hand, and help yourself."

A Perfect Nose Job

Ancient Indians mastered many surgical techniques. Based on an astonishing array of surgical instruments used by doctors in ancient India preserved in contemporary science museums, historians believe that the Indian surgeons were adept at performing a wide range of surgical procedures. Among them was Susruta (also spelt Sushruta), one of the greatest Indian surgeons from antiquity, who among other things, pioneered plastic surgery.

Susruta lived probably around 200 BC. He specialized in reconstructing amputated nostrils and torn ear lobules. A major problem in any surgical graft is the death of the transplanted segment due to lack of blood flow. To conduct delicate operations on the nose and ears, Susruta developed an ingenious technique.

For grafting fresh skin over the torn or amputated nostrils, he would first mark a small leaf-like area over the forehead of the patient. He would cut out the skin layer from the top, but maintain the stalk intact near the bridge of the nose. He would then turn the sliced skin around, and stitch it over the area where the nose had been amputated. This technique assured the preservation of circulation to the graft, and allowed it to grow in its new position. The results were spectacular: A full recovery was assured within a few weeks.

To repair a torn or amputated earlobe, he would use a similar technique of slicing a piece of skin from the scalp, preserving blood flow to the skin piece from below, turn the sliced skin over, and suture it over the torn earlobe.

But where did Susruta get his patients?

In ancient India, amputation of the nose or ears was one of the most common punishments for adultery and theft. The reason for such apparently cruel punishment was simple: The guilty could never hide his or her crime.

Al-Qurashi (Ibn-An-Nafis)— A Forgotten Scholar

The Golden Age of Islamic Medicine spanned over 300 years. During this period, brilliant scholars translated classic Greek and Latin works into Arabic, Syrian, and Hebrew, and added from their own experiences and experiments. Examples of superb Islamic scholarship can be found in the writings of Rhazes (c.860-923), Avicenna (ca. 980-1037, Fig. 3), and Hali Abbas (d. 994).

In 1924, Dr. Muhyi Tatwi of Egypt presented a dissertation in Germany, concerning Ibn-an-Nafis (Al-Qurashi, in Arabic, who died 1288), a forgotten scholar from the Middle Ages.

Because of religious restrictions, Ibn-an-Nafis could not perform human dissections. However, he had done plenty of animal dissections, based on which he proposed a nearly complete working model of human circulation. He disagreed with Galen's theories and gave a convincing description of pulmonary blood flow. He denied that there was any mixing of venous and arterial blood inside the heart, and concluded that blood gets "refined with air" in the lungs, before flowing into the left side of the heart. He gave all of his experimental details in his book *Commentaries on the Cannon of Avicenna*.

Al-Qurashi was a 13th century scientist whose work remained unknown to the West for nearly 700 years. Today he is regarded as a giant among scientists.

Get Me a New Heart

The idea for heart transplantation has been around from the beginning of the 20th century. But human heart transplantation began only since the late 1960s, when Christiaan Neethling Barnard (1922-2001), a surgeon in Cape Province, South Africa, performed the first heart transplantation.

On December 3, 1967, a young woman was admitted to his hospital, dying of head injuries from a traffic accident. At this time he was caring for a dentist, Louis Washkansky, who had suffered three heart attacks and was dying from heart failure.

When Barnard proposed that only by a heart transplant the dentist had any chance for survival, Washkansky agreed, although the procedure had never been done before.

The woman's neurosurgeons reported that she was brain dead and nothing could be done to save her. Coincidentally, her blood group and other tissues matched Washkansky's, and the woman's relatives gave permission for heart donation. Within four hours of arrival of the donor heart, 55-year-old Washkansky received the perfectly healthy heart of a 24-year-old woman. The news stunned the world and young Dr. Barnard became a household name. While the procedure was successful, after 18 days, the patient died from a lung infection.

However, Barnard did not have to wait long: in January 1968, he performed the world's second heart transplantation and this time the patient lived for 18 months, longer than any other patient who received heart transplant in the months following Barnard's first experiment.

Other people who tried heart transplantation in the early 1970s did not have such good results, leading to reduced enthusiasm for the procedure until the 1980s.

Now, heart transplantation is almost a routine procedure, with over 95% survival rates.

Doctors' Fees

See the following fee structure for doctors.
Rewards for Cures:

- Treatment for a wound, broken bone, tumor of the eye, and restoring the diseased flesh: If the patient is a rich man, ten silver coins for each therapy; if the patient is the son of a rich man, five silver coins for each; and if the patient is the slave of a rich man, two silver coins for each.

- For "Animal Doctors": If a doctor of oxen or asses cures a severe wound of an ox or an ass, he shall be paid one-sixth of a silver coin. (Considering one-fifth of a silver coin equaled the daily wages of a common man, and a silver coin paid for one year's rent, the fee structure is indeed attractive.)

Now for some "malpractice" judgments and punishment for mistreatment:

- If a patient dies while curing a wound, a tumor, or a broken bone:

- o   If the patient was a rich man, the doctor shall die.

- o   If the patient was a son of a rich man, the doctor's hand will be cut off.

- o   If the patient was the slave of a rich man, the doctor shall replace the slave.

- If the doctor knocks out the teeth during treatment:

  - o   If it was a rich man's teeth, the doctor's teeth shall be knocked out.

  - o   If the patient was a slave, the doctor pays one-third of a silver coin to the slave's owner.

- If the patient loses his eyes following treatment:

  - o   The doctor's eyes shall be destroyed.

These are some of 282 codes of Emperor Hammurabi, 500 BC. This is state-controlled regulation of medical practice at the extreme.

In spite of such restrictive Codes by Hammurabi, many people chose to become doctors. Historians believe that the laws were probably not enforced very strictly.

A comical drawing from medieval times shows a doctor as an angel when the patient is in pain; as a man pursuing a cure at all costs during treatment; and as a devil when collecting the fee.

A Sanskrit poem salutes the doctor:

Oh, honorable physician, first I salute you,
For you are the brother of Yama, the God of death.
But, behold, you are even greater,
For Yama snatches away our lives,
You, sir, our lives and wealth!

Short Stories

Royal Fistula: The Great Sun King, King Louis XIV (1643-1715) had a surgery for his anal fistula: not such a scintillating story, but this minor episode impacted medical history for a long time.

To conduct surgical repair of the royal fistula, one Dr. Charles François Felix, a court surgeon, constructed a special silver blade. In 1686, the king underwent a series of operations and needed two months to recover, at which time he was declared cured. All of France celebrated the healing of the royal posterior problem, and the blade came to be known as the Royal Scalpel.

After the king was cured of his anal fistula, his courtiers wanted similar surgery performed on themselves, even though they did not suffer from fistula. If the surgeons refused, the courtiers became furious. To ward them off, some surgeons obliged to operate on the nonexistent fistulas of court dignitaries. Even those who did not undergo such operations began putting bandages over their buttocks while attending court. This was their way to show their sympathy to the king, perhaps a trick to make sure that the monarch would inquire after their health.

A group of nuns in France undertook a creative approach to honor the king and to share in his pain and suffering: They composed a cantata entitled *DieuSauvez le Roi*. For a while, many people held a popular notion that the translated words and adapted melody from the cantata became the basis for *God Save the King*, the national anthem of England. Scholars of history, however, strongly discount and dismiss this baseless story.

Napoleon's Battles and the Ambulance: For centuries, troops wounded in battle were treated only after the conclusion of the fighting, when they would be brought back to the base camp (along with the dead) for medical help.

For many, this would prove too late. The concept of taking the hospital to the battleground was first developed by Dominique-Jean Larrey (1766-1844), a surgeon in Napoleon's army.

Larrey invented "Flying Ambulances," which were horse-drawn carts equipped with instruments, supplies, and medical personnel to treat and transport the wounded even as the battle was in full swing. Knowing that medical help was available immediately, Napoleon's soldiers fought with uncommon valor and high morale. Even Napoleon himself credited most of the successes of his campaigns to Larrey's services.

The Flying Ambulance was the forerunner of the modern ambulance.

A Hole in the Head: Making holes in the skull, or trepanation, is one of the oldest surgical remedies. The operation was practiced in almost all parts of the world. Skulls dating back 12,000 years have been recovered in many riverbeds in Europe and burial sites in Peru that show holes with signs of healing. Since bone healing is slow and takes place only if the patient is living, historians believe that the holes in the skulls were made in people who were alive, and they must have survived the procedure.

Why trepanation? These procedures were an attempt to rid the patient of evil spirits that were trapped inside the head causing migraine, epilepsy, coma following head injuries, and other ailments. Trepenation might also be a sign of a religious or ritualistic procedure. Some patients endured the procedure more than once, since skulls with holes of different sizes and varying stages of healing have been recovered from ancient burial sites.

Breast-Feeding and Wet Nursing

Human babies are designed to nurse from their mothers. As in all mammals, it is breast-feeding that has saved the human species. Yet, it is not always possible to rely upon the biological mother to breast-feed her young. Therefore, ancient societies developed alternative sources for human milk.

The most common alternative was wet nursing, or hiring a different woman able to directly nurse the baby. This practice has existed in almost all cultures since ancient times. Queens and princesses routinely employed slaves to nurse their babies so that they could maintain their beauty and form.

The famous Greek doctor of the 2nd century AD, Soranus, listed many eligibility criteria before designating a lactating woman a wet nurse. He insisted that a wet nurse should be young, not ignorant or superstitious, and above all, should not be an alcoholic because the "spirit" of the alcohol goes through the breast milk and "dulls" the senses of the baby.

In America, wet nursing became popular in the 18th century, and continued well into this century. Hospitals in Chicago and other metropolitan cities hired wet nurses to arrive each day to their obstetric wards, stay for a shift or two, and feed several babies during a single visit. They were paid based on how many infants they could feed in a single visit.

As in other cities, racism was common in mid 19th century Chicago; but for wet nursing, class, color, and race were not part of the equation. Black, brown, and white women worked as wet nurses. It is almost as if everyone knew the eternal truth that underneath the skin, all human milk is white.

Sherrington's Impulse: At the beginning of the 20th century, words such as "neuron" and "synapse" were not known, and "neurotransmission" and "electroencephalogram" did not exist.

Sir Charles Sherrington (1857-18520), the "philosopher of science," was testing if he could manage to collect the impulses passing along nerve fibers "as such" at cell junctions. In 1897 he wrote: "...we are led to think that the tip of the twig of the arborescence is not continuous with, but merely in contact with the substance of the dendrite. Such a special connection of nerve cell with another might be called a synapse," coining the word synapse.

Sherrington initially thought "syndesm" would be a good word to describe the nerve cell junction, from the Greek root. But his friend Professor Verall, an Euripedean scholar, suggested "synapse" would be better since it yields better adjective forms.

The concept of a different structure for the neuronal junction influenced neurological discoveries, especially the chemical language of the nervous system. His work stimulated the birth and growth of neurochemistry and neuropharmacology, in turn leading to discoveries of drugs to treat neurologic and psychiatric disorders. Sherrington shared the 1932 Nobel Prize in medicine.

Discovery and Delay: The path from bench to bedside has rarely been easy.

In 1928, Alexander Fleming (1881-1955) discovered that a blue-green mold inhibited the growth of Staphylococcus on a plate culture. He later showed that the mold *Penicilliumnotatum* released a substance that inhibited the bacterial growth, which he named "penicillin."

However, there was little interest in chemically producing this drug, until 1938, when Howard Florey in Oxford organized a team of scientists, including Ernst Boris Chain and Norman Heatley, to undertake the production of stable penicillin. This collaboration led to the landmark penicillin trials in the early 1940s.

Graham Liggins (1926-2010) a famous physiologist in Auckland, New Zealand was studying the mechanisms of onset of labor in the early 1970s. His research led him to serendipitously discover that maternal steroid injections helped mature the lungs of the fetus preventing them from developing a nearly-fatal disease called "hyaline membrane disease" (now being called respiratory distress syndrome or RDS). Nearly a decade earlier, it was RDS that had taken the life of President John F. Kennedy's youngest son, Patrick Bouvier Kennedy, on August 9, 1963, at two days of age after his premature birth.

Clinical trials in the 1970s further confirmed Liggins's discovery.

Yet, for reasons not clear, very few obstetricians believed in antenatal steroid treatment to prevent RDS in the babies. Over the next two decades, only 20% of eligible women were being offered this therapy. To rectify this, in 1994, the National Institutes of Health in Bethesda, Maryland, USA, organized a Consensus Development Conference, which strongly endorsed antenatal steroid therapy. Over time, steroid treatment to enhance fetal lung maturation began to be accepted as a standard practice. Now, over 90% of eligible women are given this preventive treatment that has largely contributed to dramatic improvements in survival rates for premature infants.

Two Early American Politician-Pediatricians: Pediatrics is a baby among medical subspecialties. In the 19th century some doctors began to take a keen interest in the study and treatment of diseases of children.

Two colonial doctors were the exception: They took extraordinary interest in the care of children much before pediatrics assumed the designation as a medical sub-specialty. In fact, these gentlemen were not doctors—they were politicians; governors, no less.

Governor John Winthrop, the founder of Boston, and his son John Winthrop Jr., the governor of Connecticut in Colonial America, practiced medicine and treated children.

There is no record about the medical training Winthrop the senior received. There are, however, historical records indicating that in 1643, a doctor friend of his from England sent him sheets of papers with prescriptions to be used for all "sorts of fevers" and other illnesses in "the strange land" that was America. Those sheets became the source of medical knowledge for the governor to practice medicine in the Colonies.

Winthrop Jr., on the other hand, had a more formal medical instruction. Son Winthrop was interested in science. He went to Dublin, Ireland, Padua, and Venice for formal medical training. While the governor of Connecticut, he was receiving letters from people in all walks of life concerning their health problems, seeking his advice.

He used all known remedies standard for those days, including red coral, powdered ivory, resin, saffron, balsam, rhubarb, and all kinds of herbal extraction. It is unlikely that any of those worked. But considering what was known then, perhaps some of the "remedies" gave mild relief from symptoms and others worked as placebos.

The Power of Ginseng: The United States is the largest *exporter* of ginseng root for over 100 years. In 1901, the US ginseng export exceeded one million dollars, and today the ginseng trade remains a multimillion-dollar business.

What is the medical value of ginseng? A native to northern US and Canada, ginseng was considered an aphrodisiac. It was used by American Indians to make a variety of preparations, including the "love potion," which became very popular. It contained the ginseng root, gelatin, and snake meat.

According to a legend among the Meskwaki Indian women, using the potion helped secure their mates quickly and "bag their men forever." Such claims have never been verified by using contemporary research methods.

China and the Orient are the largest importers of the Wisconsin ginseng root, and most people do use it hoping it is an aphrodisiac.

The National Center for Complementary and Alternative Medicine of the NIH is funding serious scientific research to explore the value of a variety of household remedies, including ginseng. If ginseng root is proved to be an aphrodisiac, the manufacturers will jack up its price.

A Medical History Quiz

Question 1: The 1962 Nobel Prize in Physiology or Medicine was awarded for the discovery of the structure of DNA. Besides James Watson and Francis Crick, who was the third person to share this award?

3. Marcus Fleming
4. Linus Pauling
5. Rosalind Franklin
6. Walter Krebs
7. Maurice Wilkins

Question 2: Who was Alexis Carrel and what did he do?

1. The first doctor to reach the North Pole and study the effects of hypothermia on metabolism.
2. A British psychiatrist who inspired Lewis Carroll to write *Alice in Wonderland* as a tool to understand the workings of children's minds.
3. A Canadian biochemist who helped Frederick Banting to isolate insulin.
4. A French surgeon who perfected the techniques of suturing blood vessels.

5. A Scottish priest-turned physiologist who accidentally invented electroencephalography (EEG).

Question 3: Why is smallpox called the "small" pox?

1. In ancient times, people believed that calling it "small" helped to ward off evil.
2. The smallpox virus caused small, pinpoint lesions before its mutation.
3. Named after the famous smallpox patient, Queen Elizabeth I of England, whose nickname was Small Queen Beth.
4. In archaic Greek, "small" actually meant "great."
5. Smallpox epidemics occurred in small geographic regions, compared to plague, which spread globally.
6. To distinguish it from syphilis, which caused large lesions.

Question 4: Who said, "I dress him and God healed him"?

1. Florence Nightingale
2. Harvey Cushing
3. AmbroiseParé
4. The ancient Indian surgeon Susruta

Question 5: Who said, "Gentlemen, *this* is no humbug"?

1. Edward Jenner, after the first smallpox vaccination experiment.
2. Alexander Fleming, after showing the antibiotic property of penicillin.
3. John Collins Warren, after a removing a neck tumor.
4. King James, after Mesmer's demonstration of hypnosis.
5. James Lind, to a ship's crew after showing that lemon juice cured scurvy.

Question 6: When the 16th century anatomist Andreas Vesalius reported that the rib count for men and women was the same, 12 pairs each, what was the reaction?

1. He was appointed a professor of theology the following year.
2. He was excommunicated and tried as a heretic, but found not guilty by reason of insanity.
3. He was burned at the stake.
4. His former anatomy professor called him a madman and dismissed his findings as preposterous.
5. Several hundred new students joined his class, requiring the Padua University to build a new anatomy classroom.

Question 7: Why are modern blood banks called blood "banks"?

1. A surgeon thought that storing and withdrawing blood was similar to trading money in the bank.
2. The first person who donated blood and saved a fellow employee was a London banker.
3. For the first time, large-scale transfusions were carried out in the 19th century at the banks of the river Thames in England and saved hundreds of soldiers.
4. During the Second World War, Lester Banks, a New York businessman, donated one million dollars to Albert Einstein University to establish the world's first blood-collection center.
5. It is an acronym for "Blood and Nutritional Kinetic Substances," since people believed that blood increased kinetic energy.

Question 8: At Vienna's Lying-in Hospital during the 1840s, what tasks were the doctors required to perform each morning before they worked in the obstetric wards?

1. Pray in the hospital chapels for five minutes.
2. Make rounds in the pediatric wards.
3. Do anatomy dissections and perform autopsies.
4. Speak to the family members of women in labor.
5. Repeat the *Hippocratic Oath* before entering labor wards to assure the patients about their devotion to service.

Question 9: "Student's t test" is a common statistical procedure to show if the mean values from two samples obtained from an unknown population differ significantly. It is called "Student's" t test because:

1. Pierre-Simon Laplace of France discovered the test when he was still a junior student in mathematics.
2. William Gosset, a chemistry major working at the Guinness Brewery, developed the test and published his paper under the pseudonym "Student."
3. The test was so elementary that mathematics professors expected all students to understand and use the test with no difficulty.
4. Ronald Fisher, the famous British statistician, developed the test and named it in honor of two students who died fighting for England in the First World War.
5. Erasmus Student Jacobson discovered the test and chose to call it after his grandfather, a famous 19th century mathematician, who had discovered the distribution properties of ratios obtained from unknown populations.

Question 10: Professor Wilhelm Roentgen discovered x-rays for which he was awarded the first Nobel Prize in physics in 1901. Which of the following statements is *not* true about Roentgen?

1. He chose not to accept the title of German Royalty.
2. He was expelled from school for refusing to tell on a student who drew a caricature of the teacher on a blackboard.

3. He did not patent or otherwise profit from the discovery of x-rays.
4. He took the x-ray of the hand of his wife—this was the first x-ray obtained in a human.
5. He died a wealthy man.

Question 11: René Théophile Laënnec died with little financial reward for his invention. Which of the following is also true about Laënnec?

1. He was an anti-royalist.
2. He saw scores of decapitations during the French Revolution.
3. He wrote a classical treatise on smallpox.
4. The French earned enough money from the sale of stethoscopes invented by Laennec to build the *Arc de Triomphe*.
5. He was made a Surgeon in Residence for the emperor.

Question 12: Shortly after the First World War, Louis Pasteur was voted France's most beloved son. During his lifetime, however, there was controversy about his development of the rabies vaccine. Why?

1. Some clergies thought it was the will of God that patients should die from bites of rabid dogs.
2. Critics argued that Pasteur should have studied tuberculosis instead of rabies.
3. A small number of patients receiving the rabies vaccination died, probably from an allergic reaction.
4. An assistant of Pasteur claimed that he had developed the vaccine.
5. Pasteur was sued because he had not obtained an informed consent from the mother of the first patient in whom he tried the rabies vaccine.

Answers:

Question 1: E: Physicists Rosalind Franklin and Maurice Wilkins closely collaborated in London exploring DNA structure using the x-ray diffraction technique. This helped Watson and Crick to propose the double helix model of DNA in 1953. In 1956, Franklin died of cancer at age 36, precluding her from a share of the Nobel Prize—instead, Maurice Wilkins shared the 1962 honor along with Watson and Crick.

Question 2: D. Born in France in 1873, Alexis Carrel studied medicine and specialized in surgery. He emigrated to Canada in 1906 and worked in Chicago and in New York's Rockefeller Institute. It was here he perfected techniques of suturing blood vessels and carried out pioneering research related to organ transplantation—a revolutionary procedure for that time. In 1912, Carrel was awarded the Nobel Prize in Physiology or Medicine—one of the few surgeons to receive this honor. Carrel was a prolific writer, and his book *Man the Unknown*, written for the general public, was a runaway bestseller.

Question 3: F. There was nothing "small" about smallpox—but for centuries, the distinction between various skin lesions had been unclear. All pustular skin lesions were called "pox." The skin lesions of primary syphilis were always very large, and often bulbous compared to the compact lesions of smallpox. Thus, syphilis skin lesions were called the "Great Pox" or the "Large Pox" and the designation of "small" was left for smallpox.

Question 4: C. Ambroise Paré (1510-1590), one of the most celebrated French surgeons, changed the course of renaissance surgery. On one occasion, he ran out of boiling oil, a traditional treatment for pouring over post-amputation wounds and gunshot lesions. Paré improvised and used dry dressings, worrying all night, thinking that the patients would die. To his delight, all the patients recovered, and those treated with boiling oil had developed fever and suppurative lesions. Paré vowed never to use hot oil again, and he taught the same. He wrote and taught extensively on many surgical conditions. This quote is a philosophical summary of life recorded in some of his works.

Question 5: C. In 1844, Horace Wells, a Connecticut dentist, felt that nitrous oxide could act as an anesthetic and arranged for a live demonstration at the Massachusetts General Hospital. This demonstration was an utter failure, as the patient flinched with pain during a tooth removal, prompting surgeon John Collins Warren to declare that this was "humbug." Two years later, William Morton, another dentist and a student of Wells, convinced Dr. Warren to allow another demonstration of the anesthetic effect, this time using ether. In 1846, in the MGH operating theater, Dr. Warren removed a neck tumor as Morton administered ether. The patient did not experience pain; Dr. Warren then turned to his students and said: "Gentlemen, *this* is no humbug."

Question 6: D. His professors dismissed his findings. They even declared that "if men and women had the same number of ribs, it would be because of the moral decadence in recent times."

Question 7: A. Dr. Bernard Fantus came up with the idea of storing blood in refrigerators, which could later be used for emergency transfusions; he thought it was similar to saving money in the bank that can be withdrawn when needed.

Question 8: C. Doctors and students had to carry out autopsies and anatomy dissections in the morning prior to working in the labor wards.

Question 9: B. Because Guinness Brewery prevented its employees from publishing scientific papers, William Gosset, who had discovered the test, published them under a pseudonym, "Student."

Question 10: E. Roentgen was one of the few great scientists who refused to cash in from their inventions. Roentgen died a poor man.

Question 11: B. He saw scores of decapitations.

Question 12: C. Some patients died of allergy.

Medical History Tweets

Chicago's First Doctors: After Fort Dearborn closed in 1836, its surgeon Philip Maxwell became Chicago's first doctor. He was a skilled surgeon, but he also learned the trades of civilian physicians to treat malaria, typhoid, and digestive diseases. Over 20 years, Maxwell became legendary for his big size, colorful personality, and boisterous humor. Within four years of his arrival, Chicago grew from 350 to over 4,000 inhabitants, with a dozen new doctors migrating from northeastern states.

Keeping Good Notes: Lady Macbeth's sleepwalking scene (Macbeth V, i, 1) is a supreme example of a troubled subconscious mind. In this scene, a doctor notes that the lady had "great perturbation in nature" and was rubbing her hands in a gesture to cleanse them of bloodstains. As he writes his notes, the doctor remarks: "Hark, she speaks, I will set down what comes from her, to satisfy my remembrance the more strongly." The doctor admits that the case was beyond his capabilities and turns her over to a "divine" physician. Good note-keeping habit, great medical ethics in seeking expert second opinion.

Monkey Business: The practice of monkey trading dates back nearly 5,000 years. Monkeys served as pets and performed in circuses. Galen (c.157-175 AD) used them to study human anatomy. In the twentieth century primates were important animals used in laboratories the world over. By the mid-1960s, over 100,000 monkeys were imported annually to the US for research. This number later dropped to 20,000 because of restrictions. In 1925, the Carnegie Colony began the first US primate breeding colony. Now there are only a handful of such facilities.

Of Mice and Men: In the 1800s, rats were popular sports objects, used for racing and betting in Europe. Because they were gentler than their gray cousins, albino rats also served as pets. The world's first rat breeding colony began at the University of Chicago in 1893. The rat nervous system is similar to that of human's. After 75 years of research, scientists know the behavior and psychology of rats more than that of any other species, except man.

Doctor's Gold: Even in medieval times, medical training required a long and arduous course of study. Because all medical texts were in Latin, the doctor had to master Latin. Most doctors were regarded as learned men. However, their greed was also notorious and the subject of withering satire. English poet Geoffrey Chaucer (1340?-1400) wrote about a physician:

> "...he knew the cause of every malady, were it of hot or cold, of moist or dry...he was very perfect...For gold in physik is a fine cordial, therefore he loved gold exceeding all."

Monopoly Guilds: Early medical organizations were formed solely to protect doctors. For centuries, anyone claiming to be a doctor could treat and collect a fee. This led to uncontrolled quackery. In a petition to King Henry VIII, a group of doctors in 1511 claimed that the innocent public must be protected from such "cunning," and the king ought to protect them. They asked for formation of a guild to formalize education and licensing. The king agreed, and established the Royal College of Physicians of London—the first organized medical group, which survives to this day.

Colonial Medicine: The medical men who came from England to early American colonies were not from the top of the professional spectrum. Competent and prosperous doctors stayed in England with their lucrative practices. Early colonial doctors were therefore mostly lay persons with any bit of medical knowledge. Ministers of gospel also frequently practiced medicine. In Boston, Rev. Cotton Mather (1663-1728) became famous as a doctor, and even began the administration of inoculations against smallpox.

Early Medical Schools: Most colonial doctors were trained in England, because although Harvard College, the College of William and Mary (founded in the 17th century), and Yale (founded in 1701) gave liberal arts training, there were no medical schools. A young Philadelphian, Dr. John Morgan (1735-1789), founded the first US medical school in the College of Philadelphia (later the University of Pennsylvania), with just two faculty members. King's College in New York (later Columbia University) and Harvard University followed suit two years later with six faculty members each. By the time of the American Revolution, there were 48 native medical graduates in the USA.

The Ninth Mosquito: He wanted to be a poet. But as per his parents' wishes, Ronald Ross became a doctor. Researching in India, he dissected mosquitoes to see if they carried the malaria parasite. On August 20, 1897, his assistants caught ten rare kinds of mosquitoes with wing spots. In the first eight he found nothing. In the stomach of the ninth mosquito that had just bitten a volunteer, Ross found a cyst of the malaria parasite. From this he proposed his hypothesis about the life cycle of the parasite. Ross received the 1902 Nobel Prize in medicine for this discovery, and August 20th of each year is celebrated as "Mosquito Day."

Sacred No More!: The specter of an otherwise healthy man suddenly falling with a violent outburst and shaking made the ancients regard epilepsy as a divine malady. The "sacred disease," or a seizure episode, was equated to possession by a demon or a god. Hippocrates of Cos (460-370 BC) was the first to dispel this notion. In his book, *On the Sacred Disease*, he insisted that excessive "cold phlegm" from the brain blocked vessels of the head; thus epilepsy was not due to "the will of god," but resulted from natural causes.

All in the Numbers: In 1662, a London garment dealer published

The Nature and Political Observations Made Upon the Bills of Mortality, in which he analyzed the weekly reports of births and deaths in that town. This was the first attempt at quantifying disease patterns in a population. He noted higher mortality in men compared to women, and a seasonal pattern to death rates that had initially been alluded to by Hippocrates (c 460-377 BCE). He was also the first to recognize the value of collecting community health statistics to understand human illnesses. Thus, a man with no medical training contributed to the birth of "epidemiology."

Blood of My Blood: Human blood has been misunderstood ever since antiquity. Until recently it was believed that blood transfusions transferred the donor's character to the recipient. Samuel Pepys (1633-1703), the famous English diarist, wondered what would happen if a Quaker's blood were given to an Anglican. German doctors tried to patch up marriages by transfusing blood between husband and wife. To maintain her royal dignity and powers, Queen Christian of Sweden (1626-89) is said to have made it clear that, if she should ever need a transfusion, the blood had to come from a lion.

Wrong Second Opinion: In its early days, the use of anesthesia was met with a lot of criticism. Dr. John Snow (1813-1858) of cholera epidemic fame gave chloroform to Queen Victoria during childbirth. Her journal entry states that *"...the blessed chloroform* [had a] *mild effect* [on me]...*calming and beautiful beyond bounds."* However, her army surgeon Dr. John Hall preferred surgery with no anesthesia. His opinion to doctors in 1854: *"A good hand on the knife is stimulating. It is much better to hear a fellow shouting than to see him sink quietly into his grave."*

Pencil and Penicillin: The word pencil came from peniculus, or "a little tail," because horses' tails were used to make fine brushes for use by artists. This word was later adapted for the lead pencils. In 1912, a mold, the type used often in making cheese, was named penicillium notatum, because it resembled an artist's brush. In 1928, Sir Alexander Fleming (1881-1955) showed that this mold prevented bacterial growth. After a 13-year delay, an extract from this mold was prepared and tested. This was penicillin—a product made from an object that looked like a pencil.

When an Expert Errs: In 1904, a German surgeon, F. Sauerbruch developed a fantastic method to prevent the lungs from collapsing during chest surgery. A negative pressure chamber was built to keep the patient's chest inside, while the head was kept outside for the anesthesiologist's access. The chamber was large enough to allow the surgeon and his entire team inside. Such negative pressure chambers became very popular. But because of a fundamental design error, they all failed. The experts had not considered that breathing involved both inhalation *and* exhalation—not just expansion of the chest using negative pressure. To keep the lungs of patients under surgical anesthesia working, one needs alternate swings in pressure—not just chest expansion.

The Colic Stone: Could the horrible pain from bladder and kidney stones be cured by wearing a semi-precious stone? People in much of the "civilized world" believed that wearing jade was therapeutic. The Spaniards, who found large quantities of jade in Peru and Mexico, called it *piedra de ijada*, or the "stone of the colic." The French abbreviated it to *le jade*, and the English simplified it to jade. Thus, when one wears this stone, unknowingly, one is enacting an age-old ritual for a cure of bladder stones.

How "Essential" is High BP? In 1856, an expert insisted that high blood pressure was "essential" for the body to pump blood through the kidney. Therefore, for nearly a century doctors believed that having a high blood pressure was good. In 1925, two surgeons, Drs. Rowntree and Adson, accidentally damaged the sympathetic nerve fibers in a young man, which normalized the patient's high BP. Contrary to expectations, the man improved, establishing that there was nothing "essential" about having a high blood pressure.

Seeing Clearly: How can the human eyes see objects located both near and far so well? Scientists thought that the eyeballs elongated or shortened to accomplish this phenomenon called the accommodation of the eye. In 1793, Thomas Young (1773 –1829) and English polymath and physicist disproved this. He showed that accommodation occurred because certain eye muscles, rather than the entire eyeball, contracted to change the shape of the lens; the science of optics was born. To honor him for this discovery, the next year Young was elected Fellow to the Royal Society of London. This genius was still a medical student at that time.

Some Good Words for You

The study of word origins can be a rich and entertaining exercise. A brief list below shows the delightful origins of some commonly used words in medicine.

Alveolus: From the Latin word alvus, meaning a small hollow cavity. In the 16th century, Vesalius used this word to describe the socket of a tooth. Malpighi, in the 17th century, was the first to use alveoli as the term to describe the lung units.

Ambulance: From the Latin word ambulans meaning "capable of walking." Dominique-Jean Larrey (1766-1842), a surgeon in Napoleon III's army developed horse carriages equipped with medical supplies and trained persons, which he called the "Flying Ambulances" to provide first aid in the battlefield. The English borrowed the concept and called them "walking hospitals."

Aorta: From the Greek word airo meaning "lift" or "carry." The aortic arch appears to strap over the heart as if it is carrying it; thus, it seems Aristotle (384-322 BC) introduced this word.

Artery: From the Greek word arteria meaning "windpipe." At autopsies of executed prisoners, ancient Greek doctors used to see large arteries devoid of blood, probably due to continued bleeding, since the arteries do not collapse. They interpreted this to mean that arteries carried the "vital air" that escaped when the arteries were cut.

Asylum: From the original Greek word asulon, which is a

composite word from "a" or not, and "sulao," to rob or plunder. Ancient Greeks used "asulons" as temples and altars of sacred places where robberies were forbidden. But such places also provided safe refuge for fleeing robbers, debtors, slaves, and criminals. From that background, the word got an extended use to signify a refuge or hospital for the insane; the word seemed to have acquired this meaning in English in about 1500.

Capillary: From the Latin terms caput or head and pilus or hair, meaning "hair on the head." Leonardo da Vinci made early observations on capillary flow.

Carbon: Carbon comes from the Latin word carbo, meaning coal. Priestly introduced it into English in 1789 and showed that it was an element.

Malar (region, pertaining to the cheek): From the Latin word "mala" or cheek. Because the cheeks are round and red like apples (at least in some!), this term probably came from another Latin word, "malum" or apple.

Oxygen: In 1775, the French chemist Lavoisier found the newly discovered gas was an "acid producee," thus to the Greek word "oxyntos" meaning "sour, producer" (as the acid) and coined the term oxygen.

Pulse: From the Latin word pulsare, meaning pushing or beating. Although known in many ancient cultures, the Greek doctor Praxagoras (340 BC) is said to have been the first to use the term "pulse" to correlate its changes with heart troubles.

Temporal Bone (or Temporal Region): From the Latin word "tempus" or time. What is the connection between time and this bone or region? Baldness or gray hair appears first on the temporal region of the head, showing the ravaging effects of time.

Vena cava: From the Latin words vena or vein, and cavus or hollow or cavity. It was believed that all veins must contain blood and all arteries air (see above). However, in dissections, the vena cavae often seemed to be empty. Therefore, these were considered exceptions. They were the "hollow veins."

Pediatric Philately

Philately is the art and science of collecting and studying postage stamps. Medical philately is the study of postage stamps on medical topics. Countries around the world issue thousands of medical stamps each year commemorating the achievements of inventors and discoverers, celebrating the conquest of diseases, and promoting healthy lifestyles. The personalities depicted in medical stamps range widely from Aristotle to Ylppö and from Avicenna to Apgar. Similarly, the topics depicted in medical stamps range from bubonic plague to zoonotic diseases and from cholera to yellow fever. Medical philately also helps disseminate messages concerning ongoing epidemics and help raise funds for medical research or to support those with handicaps, disabilities, and special needs. Here, I will selected some stamps related to the history of medicine that celebrate major milestones in the history of pediatrics.

I obtained 18 stamps from 15 countries for this article. These were from a review of 4 medical philately monographs and 68 issues of *Scalpel and Tongs, American Journal of Medical Philately* published by the American Topical Society.

The 18 postage stamps printed as a poster were issued 1962 to 2021. Fourteen of the 18 stamps depict 17 personalities: Hippocrates,

Avicenna, Harvey, Semmelweis, Nightingale, Barton, Koch, Waal, Ylppö, Apgar, Pauling, Drew, Ehrlich, Behring, MacLeod, Best, and Banting.

The remaining 4 stamps show: celebrating the first successful cesarean section in Colombia, South America in 1844, after which both the mother and her infant survived; the discovery and the clinical uses of ultrasound; promoting children's immunization in general, and a message to eradicate measles in particular by immunization.

What are the take-home messages?

Medical topic-related postage stamps are paper ambassadors. Teachers can use them in classrooms while discussing medical conditions and highlight the historical context behind the discovery of their causes, or development of processes for preventing and curing them. Such efforts are needed because the history of medicine topics are not included in the curricula of medical, dental, or nursing schools. Including such stamps in any lecture or talk brightens one's presentation.

Even a superficial glance at these stamps will convey the message that diseases are universal that don't honor national boundaries. Tuberculosis is tuberculosis, whether it hits someone in Mumbai or Mombasa. They help us to celebrate our collective struggles and victories against diseases and ignorance. Such efforts can help us expand our vistas of knowledge by unifying medicine, arts, history, and geography into a delightful collage.

Nic Waal (1905-1960). Hailed as the "Mother of Norwegian pediatrics and adolescent psychiatry." (Norway 2024)

Robert Koch (1821-1912). Using a home-made microscope, Koch discovered the tuberculosis bacillus; awarded the 1904 Nobel Prize in Physiology or Medicine. (Guinea 1983)

Stamp celebrating the development of Test-tube baby technology and in-vitro fertilization programs that spread around the world. (Great Britain, 1999).

John JR MacLeod (1876-1935) and Fredrick Banting (1891-1941) shared the 1923 Nobel Prize for insulin discovery. Charles H. Best (1899-1978) and James B. Collip (1892-1965), who, too, made major contributions for this work, were not named in the Nobel Prize. (Guyana 2002, Kuwait 1971)

Clara Barton: (1821-1912). Nurse, teacher, reformer, activist, humanitarian. She founded the American Red Cross in 1881. (USA 1983)

Jonas Salk (1914-95). Developed the injectable form of killed inactivated polio vaccine helping to eradicate polio (Niue 2005)

Preventing COVID-19 spread. The images show: masks; hand washing soap; COVID-19 testing and vaccination; doctor visits; and SARS-2 virus (Central African Republic, 2020).

Florence Nightingale (1820–1910). Famous nurse who became the symbol of compassionate care (Transkei 1983)

Paul Ehrlich (1854-1915), an immunologist and Emil von Behring (1854-1917), a physiologist who developed diphtheria antitoxin. Both von Ehrlich and Behring shared the Nobel Prize: Ehrlich in 1908 and Behring in 1901. (Germany 1954)

Stamps are often issued for disseminating health messages, as shown in these two stamps. The Singapore (2021) celebrates the 50th anniversary of the College of Family Physicians, Singapore, which shows an infant receiving an immunization shot. The stamp from Bophuthatswana (1985) has a picture of an infant receiving the measles immunization shot with a message "STOP MEASLES".

Ignaz Semmelweis (1818-1865). Showed that hand washing prevents post-partum sepsis. His contemporaries ignored his advice. (Hungary 1987)

William Harvey (1578-1657). First to correctly describe the adult and fetal circulatory systems. (Hungary 1987)

Linus Pauling (1901-94). Discovered chemical bonds of complex structures; won two Nobel Prizes: one for chemistry in 1954 and the other for peace in 1962. (USA 2007)

Charles Drew (1904-50). Prominent African American surgeon and researcher; developed scientific processes for storage of blood that helped develop modern blood banking system. (USA 1981)

Virginia Apgar (1909-1974) Developed "the score" to assess newborn infants at birth. (USA 1994)

Celebrating the development of medical ultrasound technology and its clinical applications. Images in this stamp show: neonatal brain, fetal face, fetal heart; adult kidney vessels, adult liver and testis, and 2 ultrasound machines. (Mongolia, 2000)

Hippocrates (460 BC-370 BC), from Greece and Avicenna (980-1037), from Persia. They practiced medicine using scientific principles rather than magic or superstition. (Iran, 1962)

Arvo Ylppö (1887-1992) Founder of Finnish maternity and pediatric sciences; defined prematurity and conducted preterm birth research. (Finland, 1987)

Celebrating the first successful cesarean section in Colombia, South America, in 1844; also commemorating the first centennial of the Nacional University. This stamp is based on a mural by Enrique Grau (1920-2004.) (Colombia, 1967)

Selected Bibliography

[Information for some stories are scattered in different sources cited below; and the General section citations provide source materials for many stories in this book.]

General

Bender G, Thom RA. *Great Moments in Medicine.* (Detroit, 1966; Park Davis & Co, Northwood Institute Press)

Bettmann OL. *A Pictorial History of Medicine.* (Springfield, IL. 1956; Charles Thomas)

Bynum WF, Porter R. *Companion Encyclopedia of the History of Medicine, Volume 1, and Volume 2.* (London, 1993; Routledge)

DeBono E. *Eureka! An Illustrated History of Inventions from the Wheel to the Computer.* (New York, 1974; Holt, Rinehart and Winston)

Garrison FH. *An Introduction to the History of Medicine.*4th Edition. (Philadelphia, 1929; W. B. Saunders)

Greenburg A. *From Alchemy to Chemistry in Picture and Story.* (Hoboken, NJ, 2007; Wiley-Interscience, John Wiley & Sons)

Haggard HW. *The Lame, the Halt, and the Blind: The Vital Role of Medicine in the History of Civilization.* (New York, 1932; Harper & Brothers)

Lyons AS, Petrucelli RJ. *Medicine, an Illustrated History.* (New York, 1978; Harry N. Abrams Inc., Publishers)

Magill FN. (ed). *The Nobel Prize Winners: Physiology or Medicine.* Volumes 1, Volume 2, and Volume 3. (Pasadena, CA, 1991, Salem Press)

Magner LN. *A History of the Life Sciences.* 2nd Edition. (New York. 1994; Marcel Decker)

Mantin P, Pulley R. *Medicine Through the Ages.* (The Bath, Great Britain, 1988; Stanely Thornes Publishers)

Martí-Ibáñez F. *Centaur: Essays on the History of Medical Ideas.* (New York, 1958; MD Publications)

Miller J. *The Body in Question.* (New York, 1978; Random House)

Porter R. *Cambridge Illustrated History of Medicine.* (Cambridge, 1996; Cambridge University Press)

Raju TNK. *Nobel Chronicles: A Handbook of Nobel Prizes in Physiology or Medicine.* (Indianapolis, 2002; Author-House)

Robinson V. *The Story of Medicine from Medicine Man to Modern Physician.* (New York, 1931; Tudor Publishing)

Wilbur CK. *Antique Medical Instruments.* (Atglen, PA, 1987; Schiffer Publishing)

Mighty Microbes

Franzen C. Syphilis in composers and musicians—Mozart, Beethoven, Paganini, Schubert, Schumann, Smetana. *Eur J Clin Microbiol Infect Dis* 2008: 27;1151—7

Nixon, JA. British prisoners released by Napoleon at Jenner's Request. *Proc Royal Soc of Med* 1939;32:877—83

Mombouli J-V, Ostroft SM. The remaining smallpox stocks: the healthiest outcome. *Lancet* 2012, 379;10—12

Raju, TNK. *Lancet,* 1998, 351;843

Rothschild BM. History of syphilis. *Clinical Infections Diseases.* 2005, 40:1454—63

Rothschild BM. On the antiquity of Trepenemal infection. *Medical Hypotheses*. 1989; 28:181—4

Tognotti E. The rise and fall of syphilis in renaissance. Europe. *J Med Humanit* 2009; 30:99—103

World Health Organization. Smallpox. The WHO webs ite http://www.who.int/mediacentre/factsheets/smallpox visited, November 24, 2011

World TB Day. http://www.worldtbday.org/ visited November 24, 2011

Parts and Principals

Abeloos M, Barcroft J, Cordero N, Harrison TR, Sendroy J.The measurement of the oxygen capacity of haemoglobin. *J Physiol.* 1928;66:262—266

Barcroft J, Barron DH, Cowie AT, Forsham PH. The oxygen supply of the foetal brain of the sheep and the effect of asphyxia on foetal respiratory movement. *J Physiol.* 1940;97:338–346

Barcroft JB. *Researches on Pre-natal Life*. Oxford, England: Blackwell Scientific Publications; 1946

Comroe JH, Jr. *Exploring the Heart: Discoveries in Heart Disease and High Blood Pressure*. (New York, 1983; W. W. Norton)

Delmonico FL. The implications of Istanbul: Declaration on organ trafficking and transplant tourism. *Current Opinion in Organ Transplantation* 2009; 14:116–119.

Dunn PM. Sir Joseph Barcroft of Cambridge (1872–1947) and prenatal research. Arch Dis Child Fetal Neonatal Ed. 2000;82: F75–F76

Echolas H. *Operators and Promoters: The Story of Molecular Biology and its Creators*. (Berkeley, CA. 2001; University of California Press)

Gartler SM. The chromosome number in humans: a brief history. *Nature Reviews Genetics*. 2006; 7:655—660.

Henio RM. *The Monk in the Garden: The Lost and Found Genius of*

Gregor Mendel, the Father of Genetics. (Boston, 2000; Houghton Mifflin Company)

Human Genome Project Information. The United States Federal Government Website. http://www.ornl.gov/sci/techresources/Human_Genome/project/timeline.shtml visited January 10, 2012.

Lee B, Davidson BL. Gene therapy grows into young adulthood: special review issue. *Human Molecular Genetics,* 2011: Review Issue 1:10;R1.

Kail AC. *The Medical Mind of Shakespeare.* (Balgowlah, NSW, Australia, 1986; Williams and Wilkins)

Mathison R. *The Eternal Search: A Kaleidoscopic Survey of the Lore and Legend of the World's Drugs and Medicines.* (New York, 1958; G. P. Putnam's Sons)

Matthews PM, McQuain J. *The Bard on the Brain. Understanding the Mind Through the Art of Shakespeare and the Science of Brain Imaging.* (New York 2003; Dana Press)

Raju TNK: The history of neonatal respiration: Tales of heroism and desperation. *Clinics in Perinatology.* 1999; 26:629–40

Raju TNK: The principles of life: Highlights from the history of pulmonary physiology. In Donn SM (ed): *Neonatal and Pediatric Pulmonary Graphics: Principles and Clinical Applications.* Armonk, NY, Futura Publishing, 1998, p 3

Reliefs and Remedies

Huang J, Mao Y, Millis JM. Government policy and organ transplantation in China. *Lancet* 2008; 278: 1937-38

Ortega R. *Written in Granite: An illustrated history of the Ether Monument.* Boston, 2006; Plexus Management

Rothman DJ. Ethical and social consequences of selling a kidney. *JAMA* 288 (13):1640–1.

Storng, GF. Pick's Disease (Mediastino-pericarditic Pseudo-cirrhosis of the Liver): A Case, with Pericardial Resection and Recovery. Can Med Assoc J. 1938 Sep;39:247-9

Young JH. *The Medical Messiahs: A Social History of Health Quackery in Twentieth-Century America.* (Princeton, NJ, 1967; Princeton University Press)

For Babies and Moms

Accardo P. William John Little and cerebral palsy in the nineteenth century. *Journal of the History of Medicine.* 1989; 44:56—71

Baker JP. The Machine in the Nursery: Incubator Technology and the Origins of Newborn Intensive Care (Johns Hopkins Studies in the History of Technology) Baltimore, 1996, Johns Hopkins University Press.

Beller FK. The cerebral palsy story: A catastrophic misunderstanding in obstetrics. *Obstetrics and Gynecology Survey.* 1995; 50:83

Blumenfeld-Kosinksi R. *Not of Woman Born: Representations of Caesarean birth in Medieval and Renaissance Culture.* (Ithaca, NY, 1990; Cornell University Press)

Cameron HC. Spasticity and the Intellect: Dr. Little versus the obstetricians. *Cerebral Palsy Bulletin,* 1(2); 1958, page 1—5

Cone TE Jr. *History of American Pediatrics.* Boston, Little Brown, 1979, p 57

Isaiah AB, Sharfman B: *The Pentatueuch and Rashi's Commentary: A Linear Translation into English.* Exodus. New York, S.S.& R. Publishing, 1960, p 9

Little WJ.On the influence of abnormal parturition, difficult labours, premature births and asphyxia neonatorum on the mental and physical condition of the child, especially in relation to deformities. *Transc Obstet. Soc London.*1861-62; 3:293

Neonatology on the Web—website providing up-to-date information on neonatal medicine, as well as other interesting items related to the history of neonatal medicine. http://neonatology.org/neo.diversions.html Retrieved, January 16, 2012.

Noakes TD, Borresen J, Hew-Butler T, Lambert MI, Jordaan E.

Semmelweis and the aetiology of puerperal sepsis160 years on: an historical review. *Epidemiol Infect.* 2008;136, 1–9

Nuland SB. *The Doctors' Plague: Germs, Childbed Fever, and the Strange Story of Ignac Semmelweis.* New York, NY. WW Norton & Co. 2003

O'Dowd MJ, Philipp AE: The History of Obstetrics and Gynecology. New York, Parthenon, 1994

Osler W. *The Cerebral Palsies of Children. A Clinical Study from the Infirmary for Nervous Diseases.* Philadelphia. P. Blakiston. 1889. [Facsimile reprinted in: *Classics in Developmental Medicine* Oxford, England. Blackwell Scientific. No 1. 1987]

Raju TNK: Some famous "high-risk" newborn babies. In Smith GF, Vidyasagar D (eds): *Historical Review and Recent Advances in Neonatal and Perinatal Medicine. Vol 2, Perinatal Medicine,* Evansville, IN, Ross Publication, 1984, p 187

Semmelweis, I. *The Etiology, the Concept and the Prophylaxis of Childbed Fever together with the "Open Letters."* (Birmingham, 1981; The Classics of Medicine Library)

Sterne L: The Life and Opinions of Tristram Shandy, Gentleman. New York, Penguin, 1997, p231

Tempkin O. *Soranus' Gynecology.* (Baltimore, 1956; Johns Hopkins Press)

Wertz RW, Wertz DC. *Lying-In: A History of Childbirth in America.* (London, 1977; The Free Press, Collier Macmillan)

Young JH: *Caesarean Section, The History and Development of the Operation From Earliest Times.* London, H.K. Lewis, 1994

Incredible Experiments

Bonner TN. *Medicine in Chicago: 1850-1950.* 2nd Edition. (Urbana, IL, 1991; University of Illinois Press)

Forssmann-Falck R. Werner Forssmann: A Pioneer of Cardiology. *Am J Cardiol* 1997;79:651—660

Raju TNK. Hot Brains: Manipulating body heat to save the brain. *Pediatrics* 2006; 117:2320—1

Seigel, D. Werner Forssmann and the Nazis. (Letter, and a reply) 1997;80:1643—44

Shorter E. *A History of Psychiatry: From the Era of the Asylum to the Age of Prozac*. New York, John Wiley & Sons. 1997; 190—239

Smith R. Julius Wagner-Jauregg: 1927. In, Magill, FN (ed); *Nobel Prize Winners Physiology or Medicine*, Volume 1. Pasadena, CA., Salem Press, 1991, 275—284

Thorwald J. *The Triumph of Surgery*. (Binghampton, NY, 1957; Pantheon)

Valenstein, ES. *Great and Desperate Cures: The Rise and Decline of Psychosurgery and Other Radical Treatment for Mental Illness*. New York, NY. Basic Books, 1986

Winthrobe MM: *Blood: Pure and Eloquent*. New York, McGraw-Hill, 1981

Mortals and Martyrs

Bliss M. William Osler: *A Life in Medicine* (Oxford, 1999; Oxford University Press).

Goldstone L, Goldstone N. Out of the Flames. *The Remarkable Story of a Fearless Scholar, a Fatal Heresy, and One of the Rarest Books in the World*. (New York, 2002; Broadway Books)

McAlister NH. John Hunter and the Irish giant. *Canadian Medical Association Journal*, 1974; 111:256-7.

National Registered Historic Landmarks. Williams, Daniel Hale, House. http://tps.cr.nps.gov/nhl/detail.cfm?ResourceId=1537& ResourceType=Building Retrieved, December 3, 2011.

Nuland S. B. *Doctors: The Biography of Medicine*. Second Edition. (Vintage Books, New York, 1995)

Plackett RL. *'Student': A Statistical Biography of William Sealy Gosset*. (Oxford, 1990; Valrendon Press).

Raju TNK. William Sealy Gosset and William A. Silverman: two "students" of science". *Pediatrics;* 116 (3): 732–5.

The Royal College of Surgeons webpage: http://www.rcseng.ac.uk/museums Retrieved December, 1, 2011.

Saunders J.B.deC, and O'Malley C.D. *The Illustrations from the Works of Andreas Vesalius of Brussels.* (Dover Publication Inc. New York, 1950)

Miscellaneous

Bonner TN. *Medicine in Chicago: 1850-1950.* 2nd Edition. (Urbana, IL, 1991; University of Illinois Press)

Byrne R. In: *1911 Best Things Anybody Ever Said.* New York, NY. Fawcett, Columbine. 1988, Page 128

Cassedy JH, Jr. *Medicine in America: A Short History.* (Baltimore, 1991; Johns Hopkins University Press)

Finn R, Orlans DA, Davenport G.A much-misunderstood caduceus and the case for an Aesculapion. *Lancet,* 353; 1999, 1978

Gluckman, L. The caduceus—a further interpretation. *New Zealand Medical Journal.* July 1998; 281—82

Henderson, J. The great snake debate continues. *Bulletin of the American College of Surgeons.* 77; 1992; p 37—38

Joubert L. *Popular Errors.* (Tuscaloosa , Alabama, 1989; University of Alabama Press)

Rakel, RE.One snake or two? *JAMA* 253; 2369; 1985

Ring ME. *Dentistry: An Illustrated History.*(New York, 1982; Aberdale Press, Harry N. Abrams Inc.)

Schouten J. *The Rod and Serpent of Aesculapius: Symbol of Medicine.* Amsterdam, Elsevier, 1967: p 118—9

Wain, H. *The Story Behind the Word.* Springfield, Il. Charles C. Thomas Publishers. 1958. Freeman, MS. The Story Behind the Word (Professional Writing Series) Philadelphia, ISI Publishers. 1958

www.ingramcontent.com/pod-product-compliance
Lightning Source LLC
Chambersburg PA
CBHW040753220326
41597CB00029BA/4759